FAST-CASUAL CHAIN

(美)瓦莱丽·克利弗/编 鄢格/译

连锁餐厅

辽宁科学技术出版社

Contents 目录

Interior and Architectural Concepts for the Fast-Casual Chain
连锁快餐厅建筑室内设计理念

006 *Branding*
品牌标识

014 *Site Selection*
选址

024 *Architectural Design*
建筑设计

028 *Interior Design*
室内设计

038 *Consider the Queue*
排队因素

Case Studies
案例分析

042 *McDonald's*
麦当劳
McDonald Concept Racine, Villefranche-de-Lauragais(31), France
麦当劳概念店，洛拉盖自由城，法国
McDonald's Graz Jakominiplaza Instore, Styria, Austria
麦当劳格拉茨 Jakominiplaza 店，斯蒂里亚，奥地利
McDonald's Haus im Ennstal, Austria
麦当劳豪斯恩斯河谷地区店，奥地利

064 *Burger King*
汉堡王
Burger King, Goldhill Centre, Singapore
汉堡王金岭中心店，新加坡

072 *Bembos*
Bembos 素食汉堡店

Bembos La Fontana, Perú
Bembos 素食汉堡拉丰塔纳店，秘鲁

Bembos Larco, Perú
Bembos 素食汉堡拉戈店，秘鲁

Nando's 86
Nando's 快餐厅

Nando's Dublin, Ireland
Nando's 都柏林店，爱尔兰

Nando's Ashford, UK
Nando's 阿什福德店，英国

Nando's Canberra, Australia
Nando's 堪培拉店，澳大利亚

Pizza Hut 110
必胜客

Pizza Hut Prime Time, UK
黄金时段 — 必胜客，英国

Pizza Express 118
PizzaExpress 快餐厅

Living Lab, UK
体验实验室——PizzaExpress 快餐厅里士满店，英国

PizzaExpress Plymouth, UK
PizzaExpress 普利茅斯分店，英国

Jamie's Italian 134
Jamie's Italian 快餐厅

Jamie's Italian, Westfield, UK
Jamie's 意大利风格餐厅韦斯特菲尔德购物中心店，英国

Mangiare 144
Mangiare 快餐厅

Mangiare Spitalfields, UK
Mangiare 快餐斯皮特菲兹店，英国

Mangiare London, UK
Mangiare 快餐伦敦店，英国

154 *Cocos*
Cocos 快餐厅
Cocos Parndorf, Austria
Cocos 帕斯多夫店，奥地利
Cocos Passau, Germany
Cocos 帕绍店，德国

170 *Sosushi*
Sosushi 寿司店
Sosushi Rho, Italy
Sosushi 寿司店罗镇分店，意大利
Sosushi Train Turin, Italy
Sosushi 寿司店都灵火车站分店，意大利
Sosushi Sassari, Italy
Sosushi 寿司萨萨里店，意大利

188 *Costa Coffee*
咖世家
Costa Coffee, Great Portland Street, UK
咖世家大波特兰街店，英国

196 *Yoshinoya*
吉野家
Yoshinoya TM, Hong Kong, China
吉野家荃湾店，香港，中国
Yoshinoya Mongkok, Hong Kong, China
吉野家旺角店，香港，中国

210 *Technical Guidelines*
技术标准

222 *Index*
索引

Interior and Architectural Concepts for the Fast-Casual Chain
连锁快餐厅建筑室内设计理念

When it comes to the creation of a successful fast-casual restaurant, savvy restaurant operators know that every aspect — from concept development to site selection to build-out and design — must work together. That does not mean there are not hurdles on the road to opening day, but the more organized the operator, the better chance they have for success.

若要成功打造一个连锁快餐厅，睿智的经营者应该熟知其涉及的每个方面，从选址到建造再到设计，所有的环节必须融合在一起。即便如此，这并不意味着从最初构思到餐厅营业能够一帆风顺，但是越是精心策划，成功的几率就会越高。

McDonald Concept Racine in Villefranche-de-Lauragais, France　麦当劳概念店，法国洛拉盖自由城

Chapter 1: Branding

Building your brand

Fast-casual interior and architectural design is about more than what patterns to choose for the chairs or the type of slate rock needed for the retention wall. This guide covers everything that new or existing restaurateurs need to know in order to build and grow their brands from the ground up. After all, in today's competitive restaurant environment, fast casuals are poised to experience startling success, and the well-designed brand will garner a great amount of consumer interest and sales.

To be successful, the best fast-casual brands have found their niche and stuck with it. Whether an operator is looking to capture the bakery café, Asian, fresh-Mex or burger market, a well-defined brand can mean the difference between one that flourishes and one that founders.

When looking to expand an existing operation or build one from scratch, it is important for operators to define their brand and what they want it to mean to their guests.

'Even the most successful brands must constantly adapt to keep pace with cultural changes and shifting consumer preferences,' said Duke Marketing founder and president Linda Duke. 'Owning a powerful brand enables you to capture and retain consumer loyalty, and provides the leverage and credibility to expand your brand into new markets and categories and to introduce new products.'

Finding the right niche is a key element in establishing a strong and lasting concept.

'In Texas, opening a Mexican restaurant would be hard because of the established chain restaurants and mom and pops. It is important to know your market — what is existing, what is working for the operators already there and what is a way to differentiate yourself,' said Paul Munsterman, president of Dallas-based Monster Design. 'There are ways to spin off

建立自己的品牌

连锁快餐厅的建筑室内设计不仅仅是选择何种样式的座椅或者采用何种规格的板岩建造护土墙。这一指导准则涵盖多个方面，有助于餐厅经营者打造属于自己的特色品牌，在众多餐厅中脱颖而出。总之，在现今餐厅经营竞争日趋激烈的市场环境下，连锁快餐厅仍然能够蓬勃发展，精心策划的品牌形象毋庸置疑带来了极大的帮助。

那些突出的品牌餐厅已经找到属于自己的领地，并取得成功。对于一个经营者来说，无论选择进入何种市场，面包烘焙、亚洲餐饮、墨西哥餐饮或者汉堡制作，一个明确定位并精心策划的品牌往往能够起到至关重要的作用。

无论将原有餐厅进行扩建或者打造一家新的餐厅，经营者首先要进行品牌定位，并清晰地向顾客传递出其多蕴含的意义。

杜克营销传播公司负责人，琳达·杜克（Duke Marketing，加州一家整合营销传播公司），曾经说过："即便是非常成功的品牌也需要不断地改进，以去迎合不断变化的文化及顾客喜好。拥有一个成功的品牌可以帮你留住顾客，为进军新的市场和领域提供信誉保证，便于推出新的产品。"

市场定位是建立有力而持久品牌的重要元素。

保罗·穆斯特曼（Paul Munsterman）达拉斯一家营销机构（Monster Design）负责人这样说："要想在得克萨斯州成功开设一家墨西哥餐厅一定是非常困难的，因为这里的文化和已有的连锁餐厅类别，

into different directions rather than go straight into the competition.'

Restaurateurs must perform a solid amount of brand analysis and due diligence to identify the specific message they want to convey and the type of niche they want to fill. Part of that research includes a demographic and psychographic analysis and a competitive breakdown of existing brands in local and regional markets. Part of the research should include income levels for the markets the operator is looking to reach.

Your brand should be derived from who you are, who you want to be, and who people perceive you to be.

'Checking what the average income is in the area will help to dictate how much you should spend in terms of design and architecture,' said Michelle Bushey, creative director and partner with Dallas-based Vision 360 Design. 'You do not want to design yourself out of your neighbourhood or surroundings.'

Bushey worked on the concept development of Dallas-based Mooyah Burgers & Fries and said having a good business plan for any concept plays a huge role in its developmental success.

'Creating a plan in which to work from takes time and research,' she said. 'Your plan should provide breakdowns of costs and expenses, which will help determine the amount of business you need to generate to both break even and make a profit. Some clients use market research firms, but if you do not have that in your budget you can start putting the information together yourself through your local chamber of commerce or commercial real estate brokers.'

Almost every community has a chamber of commerce, which can provide demographic and real estate breakdowns of the region an owner or operator is trying to reach. More research can be done at local libraries and by visiting regional and national restaurant association Web sites.

对于餐厅经营者来说，一定要熟知即将进入的市场——市场中已存在的条件、对自己的影响以及如何使自己脱颖而出。相对于直接进入一个竞争激烈的市场，转变进入方式更为重要。"

餐厅经营者必须进行一定的品牌分析，明确确认想要传达给顾客的信息以及自己的定位。其中一部分调查包括餐厅所处地区的人口和心理特质分析、当地及区域市场已有品牌分析。除此之外，还应包括市场人群收入水平。

品牌设计源于"你是谁，你想成为谁，你在别人眼里是谁"

米歇尔·布希（Michelle Bushey），Vision 360 Design 设计公司创意总监及合伙人，曾经说过："了解一个地区的平均收入可以帮助你计算出你在建筑和设计方面的预算，不要超过周围其他餐厅。"

布希在达拉斯一家名为 Mooyah Burgers & Fries 的汉堡快餐厅制定品牌发展规划时说："一个合理的经营计划在未来发展中起到至关重要的作用。"

她还提到："制定一个可行的规划需要花费时间深入调查研究。规划中应该提供各种成本数据，以便于计算出如何经营以保持收支平衡或者获得利润。一些经营者委托市场调查公司作出规划，但是如果没有这方面的预算，也可以从当地商会或者商用房地产经纪人收集数据信息，自己制定计划。"

基本上每一个社区都有自己的商会，可以提供该地区的人口统计及房地产方面信息。当然，也可以到当地图书馆或者国家餐厅协会网站查找相关信息。

Chapter 1: Branding

Relating to your customer

Once you have defined your brand, it is important to note how your brand will relate to the customer.

'A lot of people are focused on the brand from the outside in as one way to relate to the customer visually, but there's also the other side, which is looking at the brand from the inside out,' Munsterman said.

Munsterman is referring to the brand message that restaurant employees can relay to the customer and how those staff members project and understand the brand. 'It is really educating those employees on what the brand message is,' he said. 'I think a lot of places do not focus on that enough. If the staff does not understand how to portray the brand message to the customer, I think there is a missed opportunity.'

Studying the area's demographics and establishing focus groups are other ways operators can connect with and understand their guests. Comment cards also can be useful.

'It's always a surprise to hear what people have to say. Then you can refine and define your brand based on that information,' Munsterman said. 'If you are a start-up, there is much more strength in knowing what you are doing from the beginning. Friends and family focus groups work to a degree, but getting people into a room that fit right into your demographic really makes a difference.'

Operators also can connect with their guests outside the walls of the restaurant. By supporting local fund-raising events or school athletic teams, operators can further bond with the members of their community, build relationships and increase their foot-print.

Visually, murals, photos and memorabilia of the community and other artwork can connect with guests on an aesthetic level.

与顾客建立联系

品牌确定之后，应在其与顾客之间建立联系。

穆斯特曼提出："许多人习惯将品牌从外向内与顾客之间建立联系，这是其中的一种方式。其实，从内向外也可以是一种方式。"

穆斯特曼上面提到的是指餐厅员工如何理解并向顾客传达的品牌信息。"这其实是在教会员工理解什么是品牌信息，很多餐厅在这方面做得都不够。如果员工不知道如何向顾客传递信息，那么我想就错失了一个机会。"

对于餐厅经营者来说，研究当地的人口因素并确立目标顾客群是了解并与顾客建立联系的另一种方式。意见回馈卡也能起到一定的作用。

"大家的建议经常能够带来惊喜，可以根据他们的意见完善和确立自己的品牌。"穆斯特曼如是说。"如果是初次尝试，在最初就明确自己的目标可以带来更多的便利。以朋友群和家庭为调查对象可以取得一定程度的效果，但是如果能够把很多人聚集在一起则会收到更大的成效。"

当然，经营者还可以把大家聚集到餐厅外面，例如举办当地募捐活动或者学校运动会等，也不失为与潜在顾客建立联系的好方式。

另外，也可以通过社区内的墙画、图片和纪念品来与顾客群在艺术层面上建立一定的联系。

Franchisees with Jacksonville, Fla.-based Firehouse Subs are charged with commissioning a painted mural at each location, representing the local community and its fire department. Restaurant franchisees also are encouraged to hang photos and memorabilia purchased or donated by fire departments in their communities.

Firehouse Subs was founded nearly 14 years ago by firefighting brothers Robin and Chris Sorensen. The fire-house theme runs through each of the chain's more than 300 locations.

'It is a great way to build community rapport,' said Don Fox, Firehouse Subs' chief operating officer. 'Particularly when we go into a new market, it is a great opportunity for the franchisee and area rep to reach out into the community.'

Part of the chain's brand touch point is a condiment station filled with bottles of different hot sauces — an idea that was launched by chain guests.

The first restaurant did not offer hot sauces, so customers started bringing in their own and leaving them on the counter for later use.

'They would start to accumulate at the restaurant,' Fox said. 'They became conversation pieces and an opportunity to engage with and talk to the guest. Now, we have up to 50 hot sauce bottles.'

The hot sauces are complimentary and serve as another way the brand connects with its guests.

Build your brand image: from logo to menu development

Branding elements can come in a variety of shapes and sizes, but those elements need to extend throughout the entire chain and dining experience. But before the groundwork for those brand elements is laid out, a restaurant name and logo design must be developed.

Firehouse Subs（美国知名连锁快餐厅）的每一家餐厅内都有一幅代表当地社区和消防部的墙画。除此之外，餐厅内还可以悬挂社区消防部购买或者捐赠的各种图片。

14年前，Firehouse Subs 由消防队员罗宾·索伦森和克里斯·索伦森兄弟二人成立，"消防"这一主题贯穿于300多家分店内。

丹·福克斯（Firehouse Subs 的业务总裁）提到："与社区建立联系是一种特别有效的方式，尤其是在进入新的市场时，这会带来很多的机会。"

品牌的触点即为一个调味品台，上面摆满了各种盛满热辣酱油的瓶子，这个想法是由顾客提出的。Firehouse Subs 第一家店没有提供热辣酱油，所以顾客开始自己带到餐厅来，然后就留在那里供后来的顾客使用。

"然后这些调味酱就开始在餐厅聚集起来"，福克斯说到，"这些瓶子成为了谈话的话题，增加了我们和顾客交谈的机会，现在我们总共有50瓶由顾客提供的调味酱。"

这些热辣酱油同样已经成为了一种与顾客建立联系的方式。

建立自己的品牌形象：从标识到菜单

品牌元素可以以多种形式和规模呈现出来，但其必须渗透到每一家店内和每一次就餐经历中。在品牌元素运用之前，首先要确立餐厅名称和标识。

Chapter 1: Branding

'When thinking of names, we approach it very broadly at first. The ones that would appeal to us would be those that lend themselves graphically, but the sound and tone would connect more emotionally with customers', Munsterman said.

Once a name has been chosen, work can be done on the logo.

'A well-designed logo should be simple, compelling, memorable, relevant and versatile,' said Bushey.

'What you do with print, menu and logo development needs to relate directly to the food and design of the brand,' she said. 'It is a package deal. Your brand identity should be cohesive from start to finish.'

And from conception through execution, design — logo, signage or otherwise — should create solutions that can actually be implemented.

For example, Firehouse Subs easily depicts the brand message, and Daphne's Greek Café lets consumers know what type of restaurant they are visiting and the type of experience they should expect.

Munsterman said there are three main areas where the key brand elements should be implemented throughout a fast casual:

1. Signage
2. Interior design
3. Menu boards and the POS

Signage includes any outdoor message that could lead potential customers through the front door. The signage also should fit in with the type of brand message an operator is trying to convey.

Munsterman said when his company worked on the brand development of a restaurant in the Dallas region, 'but the sign fell way short. The materials just did not match what they could have done.'

"我们在取名字的时候，首先确定了一个很广的范围。吸引我们的那些名字在视觉上要图形感十足，在声音和声调上要与顾客建立一定的精神联系。"穆斯特曼解释说。

第一章　品牌标识

One way to take a brand message from the exterior to the interior is to use design elements such as murals, framed artwork or other artifacts.

举两个例子，Firehouse Subs 可以很容易地传达出品牌信息，Daphne's Greek Café（Daphne's 希腊餐厅）可以让顾客清楚地了解到他们就餐的餐厅类别以及能够带来的体验。

品牌确定之后，就可以着手标识设计。

布希提到："一个精心设计的标识应该是简洁有力的、引人注目的、令人难忘的、相关性强并且多样化的。"

她还解释说："关于印刷品、菜单和标识还应与餐厅食物和品牌设计紧密联系。这是一个整体，品牌特色从头到尾都要体现出来。"

从构思到实施再到设计，要确保找到合适的方式使标识、标牌或者其他元素都能够得到实际的应用。

穆斯特曼指出，一个快餐厅内其中有三个地方必须体现出主要品牌元素：

1. 标牌
2. 室内设计
3. 菜单板和收银台

其中，标牌主要在餐厅外传递信息并能够吸引顾客的到来。标牌应该与餐厅传递的品牌信息保持一致。在为达拉斯一家餐厅做品牌发展规划时，穆斯特曼曾说过："这个设计并不合乎标准，元素之间没有实现融合。"

将品牌标识由外向内传递的一种方式即为运用设计元素，如壁画、装裱的艺术作品或者其他工艺品。

Figure 1. Logo of Burger King
Figure 2. Nando's Ashford in UK

图1. 汉堡王标识
图2. Nando's阿什福德店,英国

Chapter 1: Branding

For example, Denver-based Chipotle Mexican Grill's clean logo and exterior design lends itself to the urban-style décor used in the restaurants' interior. Rubio's Fresh Mexican Grill, based in Carlsbad, Calif., uses a 'brand wall' at each of its locations to convey the chain's 25-year heritage through timeline photos. And Sedona, Ariz.-based Wildflower Bread Company's use of hand-blown glass designs creates a unique feel at each of the company's locations, without changing the brand image.

Interior visual elements also include the menu board and point-of-sale collateral, along with menu-item names.

The Grand Traverse Pie Company, in Traverse City, Mich., has menu items named after different areas surrounding Michigan lakes. For example, the Grilled Manitou sandwich is named after Michigan's Manitou Islands, and the Old Mission Cherry Pie is named after the Old Mission Peninsula in Northwest Michigan.

Grand Traverse is not the only fast casual known for its unique names. Moe's Southwest Grill's menu includes pop-culture-named menu items such as Joey Bag of Donuts (from 'My Cousin Vinny'); Art Vandalay (from 'Seinfeld'); and the Billy Barou (from 'Caddyshack'). Firehouse Subs also has menu names such as the Hook & Ladder, the Engineer and the New York Steamer.

'Naming is really important to support whatever that brand may be,' Munsterman said. 'If you have a Mexican restaurant, menu-item names and descriptions tend to not be so bland. But it depends on how traditional the concept is.'

Images also play a major role when it comes to displaying menu items. 'Really, colour, texture and fonts all play a big part on the menu board. Typically, people think of menu board as very straight and tight, but it does not have to be that way,' Munsterman said. 'Anything with a picture,

例如，位于丹佛的 Chipotle Mexican Grill（Chipotle 墨西哥餐厅），简洁的标识和室外设计有助于室内城市风格的运用；位于加州的 Rubio's Fresh Mexican Grill（Rubio's 墨西哥餐厅）运用"照片品牌墙"记录餐厅建立 25 年的历史；位于亚利桑那州的 Wildflower Bread Company（Wildflower 面包公司）采用人工吹制的玻璃结构营造出一种独特的感觉。

室内视觉元素还包括菜单板、收银台以及菜品名称。位于特拉费斯城的 Grand Traverse Pie Company（特拉费斯馅饼集团），其食品根据密西根湖周围的不同地区命名。例如，Grilled Manitou 三明治根据 Manitou Islands（曼尼图岛屿）命名，Old Mission 樱桃馅饼根据 Old Mission Peninsula（欧米新半岛）命名。

Grand Traverse 并不是唯一一个采用独特名称命名食物的快餐集团，Moe's Southwest Grill（莫氏西南烤肉店）采用流行文化为菜品命名，如根据"My Cousin Vinny"（《温妮表姐》）命名的"Joey Bag of Donuts"和根据"Caddyshack"（《疯狂高尔夫》）命名的"Billy Barou。"Firehouse Subs 的菜单中也出现了 Hook & Ladder（《钩子与梯子》）、Engineer（《工程师》）等名称。

"取名是对于品牌设计的补充与体现"，穆斯特曼说："如果你开的是一家墨西哥餐厅，那么菜单上的名字和描述就不能够过于平凡。当然，这还得取决于设计理念的传统程度。"

图案对于展示菜品名称同样起到重要的作用。

第一章　品牌标识

people are attracted to because it looks good. So pictures can really drive sales to an item.'

Images also can direct attention across the menu board so customers can look at all categories of menu items instead of one or two.

Other areas that can be used to incorporate brand elements include the restrooms and possibly a community board, where images of customers or notes from them can be displayed.

'One thing in the toilet that is easy to do is sound, whether it is music or other acoustics,' Munsterman said. 'You can also promote items through the audio without being too intrusive.'

穆斯特曼指出："颜色、质感和字体在菜单的设计上至关重要。在人们的印象中，菜单的设计应该直接而紧凑，但其实不一定必须是这样。人们经常被带有图案的东西吸引，因为图案看起来很美好。图案设计有的时候真的能够带动销售量。"

图案还可以直接吸引顾客的目光，让他们浏览整个菜单，而不是盯住其中的一两道菜。

其他可以体现标牌特色区域还包括休息室或者社区展示板，上面可以展示顾客提供的图片或者评论等。穆斯特曼指出："在卫生间中最简易的方式是播放音乐或者其他声音，甚至可以通过广播宣传餐厅的菜肴，但是要注意尺度。"

Definitions
Brand positioning: The specific niche in which the brand defines itself in the competitive environment. Positioning addresses differentiating brand attributes, user benefits and target segments, singularly or in combination.

Positioning statement: A concise, written statement of the positioning concept, conveying the essential features of the brand and its niche.

Messaging: The messages created to target specific markets. What should the brand conveyto specific target markets? What does it promise to deliver to that market? What makes it different and better?

名词解释
品牌定位：一个品牌在竞争中占据的细分市场领域。定位强调品牌与众不同的特征、用户利益、目标分割其中的一项或几项。

定位计划：一份详细精确的书面计划，内容涉及定位理念、品牌自身及细分市场的特征。

信息要素：品牌针对细分市场的所有信息，包括品牌应传达的内容、对市场的承诺以及如何使其与众不同或变得更好。

Figure 3. Brand information expressed in different forms

图3. 不同形式表现品牌信息。

Chapter 2: Site selection

Figure 1. BEMBOS in La Molina, Perú
图1. BEMBOS素食汉堡餐厅，秘鲁，拉丰塔纳

Site selection

Just as difficult as finding the right franchisee is finding the right site location.

'In this industry it is "location, location, location." When you are a start-up, "A" locations can be difficult to come by,' said Michelle Bushey, creative director and partner with Dallas-based Vision 360 Design. 'A lot of developers and landlords are looking for tenants with a history or track record to fill those leases.'

For most start-ups, 'A' locations are ideal for exposure, but the costs of going into the space will be higher. However, Bushey said the amount of exposure and traffic a restaurant receives usually offsets the costs of an 'A' location.

'Once you've established your brand, you have more opportunity to go into "B" locations and be successful,' she said. 'Don't just jump at the first opportunity — be cautious about your site selection.'

While some fast-casual operators may rely on a broker to help find locations, others rely on companies such as Pitney Bowes, Asterop and Windsor Realty to help with site selection.

Pitney Bowes' MapInfo is a fully integrated data, software and research-services firm that provides predictive location analytics. Restaurant-practice leader Brian Hill said restaurant owner/operators are, or should be, driven by four key factors when looking to build or expand their brand:

1. Household populations
2. Worker populations (such as nearby office parks)
3. Shopping areas
4. Other sales driver

餐厅选址

寻找合适的餐厅选址如同最初构思市场定位一样困难。

"在这个行业中，选址尤为重要。开设品牌第一家餐厅，很难找到'黄金地段'的选址。"米希尔·布希（达拉斯Vision 360 Design设计公司总监及合伙人）解释道，"很多开发商和业主喜欢把房子租给那些有历史记录可查的租客。"

对于品牌开创者来说，黄金地段是最佳选址，但租金往往要很高。然而，布希曾指出，黄金地段的便利交通和高曝光率可以抵消相对较高的租金价格。

她还曾提出："对于那些已有品牌的餐厅经营者来说，可以选择那些二级地段。但不要仓促决定，仔细考虑之后再做选择。"

一些餐厅经营者选择依靠房地产经纪人寻找位置，而另一些则依靠Pitney Bowes、Asterop和Windsor Realty等公司来帮助选址。

Pitney Bowes公司是一家集信息、软件和调查服务的综合性公司，提供预测选址分析服务。餐厅经营行业引导者布莱恩·希尔指出，餐厅经营者在打造或扩大品牌时应考虑四个主要因素：

1. 周围住户人口数量；
2. 周围上班族人口数量；
3. 周围购物中心；
4. 其他促销因素。

'Real estate, to the informed restaurateur, has always played an important role in the viability of a site,' Hill said. 'You could have very good information about who your customer is and the demographic, but if you don't know the dynamics of how worker populations or shopping affects that store or competition, that also affects real estate viability.'

Population base

For operators seeking a new restaurant location, the restaurant should sit in an area where their primary customers live and work. That often means tapping into demographic information about an area's average income, population trends and the dining and shopping habits of residents. 'From a site-selection standpoint, a lot of the work emanates from the demographics of the area,' said Houston Jones, president of Louisville, Ky.-based restaurant and design firm The Houston Group, in the special report, "Real Estate in Real Time."

Jones subscribes to a service offered by the Certified Commercial Investment Member Institute that provides specific demographic information for neighbourhoods across the country.

'Say I'm looking for steakhouses,' Jones said. 'I can filter my demographic information down to find out how many times people in that neighbourhood went out to eat at a steakhouse in the past six months.'

Don Fox, chief operating officer for Jacksonville, Fla.-based Firehouse Subs, said the key for them is population density — areas with a three-mile population base of 20,000 and an average income of \$35,000.

希尔指出："选址对于一个成功的餐厅经营者来说至关重要。好的选址可以帮助很好地了解客流量信息，但是如果对于周围动态客流（如附近上班人员数量或者商场顾客数量）没有很好的了解，同样会影响到餐厅的经营。"

人口基数

对于餐厅经营者来说，餐厅应选择在一个潜在顾客群生活和工作的区域内。这就要求对于周围区域人口构成、收入、饮食及购物习惯等信息进行充分的了解与掌握。

休斯顿·琼斯，路易维尔餐厅设计公司（The Houston Group，休斯顿集团）负责人在一篇名为《正确选址与正确时间》（Real Estate in Real Time）指出："从选址角度来说，许多工作都是从区域人口分析引申出来的。"

琼斯注册成为美国注册商业投资人员协会会员，可以从中获取某一特定地区的人口相关信息。

"比如我要寻找一家牛排餐厅"，琼斯解释说："我将相关数据输入之后，就可以找出这个地区的人群在过去6个月内在牛排餐厅就餐的平均次数。"

丹·福克斯，Firehouse Subs连锁餐厅运营总监提到，人口密度是一个重要因素——区域周围3英里之内人口基数应达到20000，平均收入35000美元。"在一个蓝领为主的市场内，我们同样能够取得成功，这和在一个人均收入超过80000美金的中层人群社区内没有什么区别"，福克斯指出："在开始

Chapter 2: Site selection

'We'll be just as successful in a market that is solid blue-collar as in a neighbourhood of an $80,000-plus median income,' Fox said. 'In advance of selling to an area rep, we map a market and an area we believe is viable. It is all very calculated of how the market is going to look. We're really looking for those "A" sites and where those attributes are where we think is going to lead to success.'

San Francisco-based Asterop Inc. — creator of business-intelligence technology dedicated to strategic and operational marketing — uses a discreet scoring function designed to evaluate all possible site locations in an area and then categorizes those locations on best to worst.

'What anybody wants to do is find more people of the same type that have led to the success of their stores in the past,' said Bryan Vais, Asterop's chief operating officer. 'Part of our approach to develop the basis to this is to make sure we're looking for the right sets of attributes about people in the retail landscape that are good indicators of store success.'

The company also uses a consumer-segmentation programme that divides the population into one of a set number of clusters. The system follows four subtleties of people's purchasing behaviour in three areas:

1. Personal goods and services
2. Home furnishings and equipment
3. Fast-moving consumer goods such as groceries and food choices

'Clusters let operators know what they spend on food and what type of food. Any offering of a fast-casual restaurant is going to resonate differently with these types of consumers,' Vais said.

Once it has been established that a specific, desirable demographic is present in an area, the restaurant chain can look for other drivers, particular types of businesses that attract traffic.

经营之前，我们深入研究市场并确定一个可实施的区域。对于市场未来发展要经过全面仔细的规划，我们确实需要找到那些黄金地段和那些能够指向成功的因素。

位于旧金山的Asterop公司是一家战略与实用营销公司，运用商业智能技术将一个地区的所有选址根据功能进行等级分类，类别从最好到最差。

"这些业主们想要找到的无非就是同一行业中成功的先例"，Asterop公司运营总监布莱恩·维斯解释说："我们的工作实际上就是要确保找到那些成功人士具备的特质，这是取得成功的直接因素。"

Asterop公司同样运用消费者分类程序将人群分成不同的团体，这一程序根据人们三种不同购买行为为基础：

1. 个人物品和服务；
2. 家居装饰和设备；
3. 快速消费品（杂货和食物）。

消费者分类确定之后，就可以找出餐厅的潜在顾客。当然，经营者也可以运用其他措施来吸引目标顾客群。

"一些相对高级的杂货店聚集区也可以帮助带来一定的客流量"，雷夫·金德，Baker Bros. 连锁集团特许经营总监告诉FastCasual.com（连锁快餐厅组织）："餐厅逐渐开始选址在这些地方，那些没有既定目标的人群往往喜欢在这些区域寻找就餐的场所。当然一些餐厅热衷于吸引流动的客流，而还有一些喜欢选址在购物区中，吸引更多的固定顾客。

For Dallas-based Baker Bros, which depends heavily on lunchtime service, daytime drivers such as universities, hospitals and large office buildings are essential. They also look for upscale neighbourhoods with a minimum household income of $50,000.

'Power centres like upscale grocery stores such as Whole Foods drive traffic to us,' Ralph Kinder,

Baker Bros. director of franchise development, told FastCasual.com. 'Restaurants tend to cluster around these things, because if someone doesn't have a specific destination in mind, they'll say, for example, here in Dallas, "Let's just go to the Galleria area and see what's there."' While some fast casuals look to those types of daytime drivers, others prefer shopping areas to drive customer traffic.

Shopping areas and other factors

While traditional shopping malls are alive and thriving in the United States, consumers have seen a proliferation of lifestyle centres in and around urban and suburban regions.

Lifestyle centres may offer the same type of shopping options as traditional indoor malls, but they are typically not enclosed, do not have a large anchor store, and give shoppers drive-by access to each entrance.

'I think there's pressure on malls these days,' Vais said. 'The best ones will always be in demand, but for certain malls of status and size, retailers are looking at their choices made in the past and re-examining them and whether they made the right choices. The lifestyle centre is in demand because it represents a different retail experience, too. It is a different feel about the place and creates a different shopping experience overall.'

More fast casuals are looking at lifestyle centres to place their next location. In Louisville, Ky., the Summit lifestyle centre is home to Quiznos, Qdoba, Starbucks and J. Gumbo's, in addition to retailers such as

Figure 2. Bembos Aviación — facade
图2. Bembos素食汉堡店圣伯加镇——外观

购物中心与其他因素

在美国,传统的购物中心一直蓬勃发展,与此同时,一种全新的购物形式——城市生活休闲中心也开始悄然兴起。

城市生活休闲中心同传统的室内购物中心提供相同的购物选择,但不再是封闭空间内的百货商店形式,顾客可以进入不同的店内购物。

维斯指出:"现阶段,那些传统的购物中心开始感受到压力。那些好的总会是人们需求的,但对于特定地位和规格的商场来说,经营者开始重新审视之前是否做出了正确的选择。城市生活中心代表一种不同的购物体验,因此被大量需求。"

越来越多的快餐连锁店开始选择在城市生活休闲中心落脚。例如,在路易维尔的Summit休闲中心内就入驻了Quiznos三明治餐厅、Qdoba墨西哥餐厅、

Chapter 2: Site selection

Banana Republic, Gap and J. Crew.

Brian Hill, Pitney Bowes' MapInfo restaurant-practice leader, said he has not seen any brand that uniquely goes after shopping centres. However, those that do should consider parking arrangements and the type and amount of seating a shopping location will provide.

'What are the things that are your "must haves" and deal killers?' Hill asked. 'Look at those markets and formats.'

In addition to lifestyle centres, nontraditional locations such as airports, stadiums and college campuses have been more appealing to fast-casual brands.

Penn Station East Coast Subs, Au Bon Pain, Wolfgang Puck Express and Einstein Noah Restaurant Group are just a few that have utilized airport, stadium and college campus locations to expand their brand. Camille's Sidewalk Café inked a deal with Wal-Mart Stores Inc. to place its eateries in Wal-Mart Supercentres across the United States.

'The greatest benefit is Wal-Mart's draw of millions of people each year. Foot traffic is no longer a concern...neither are hungry customers,' said David Rutkauskas, president and CEO of Camille's parent company, Beautiful Brands International.

The type of shopping or other location a fast casual goes into also should depend on the primary daypart it is trying to reach.

'If dinner is a focus of your brand, you will generate more business with a location in a shopping area due to the presence of nighttime traffic as compared to areas populated by office buildings, which historically do better lunch business,' Bushey said. 'But high traffic doesn't necessarily mean success. It depends on the kind of traffic that you're getting and when you are getting it.'

星巴克等。

布莱恩·希尔曾指出，没有一个品牌专门选择进驻购物中心，但是进驻之前应考虑购物中心的类别、能容纳的顾客以及停车位等情况。

希尔还提出要明确哪些是必须的，哪些是一定不需要的，这就需要看看市场以及当今形势。

除了城市生活休闲中心，机场、体育馆、大学校园等也越来越吸引连锁快餐店的进驻。

Penn Station East Coast Subs 餐厅、Au Bon Pain 面包店和咖啡厅、Wolfgang Puck Express 餐厅及 Einstein Noah 餐饮集团是为数不多的进驻机场、体院馆和大学校园的快餐厅。Camille's Sidewalk Café 同沃尔玛集团签订协议，将在其全美的超市中开设。

大卫·路特考卡斯，Camille's 母公司 Beautiful Brands International 主席及执行总监指出："进驻沃尔玛最大的益处就是其每年数以百万计的客流量，因此不再为客流量而大伤脑筋。"

购物中心或其他选址类型的选择取决于餐厅最初的市场定位。

"如果餐厅以晚餐为主，那么选址在购物区要比办公区合适得多。购物区夜间客流较多，而办公区对于午餐经营更有益处"，布希指出："但是高客流并不意味着一定成功，当然还要取决于目标客户群。"

第二章　选址

Corner or endcap?

In today's competitive restaurant environment, operators also have to consider the location designed to give them the most visibility. When deciding between a corner or endcap location, there are several things to consider.

'Most of our clients prefer endcaps due to visibility and the opportunity for outdoor seating. Patio space is not typically included in lease space, and therefore the operator pays no rent on it,' Bushey said. 'It provides additional seating and revenue without increasing costs.'

转角式或者天井式

在如今餐饮市场竞争激烈的环境下，经营者须考虑餐厅位置设计以便使其更具可见性。当在转角式或者半圆式之间选择时，需考虑多种情况。

"无论从可见性或者设置室外就餐区考虑，大多数经营者都喜欢天井式。天井空间一般不包括在租赁空间内，因此不需要付租金"，布希指出："经营者可以在不需要增加预算的情况下，打造额外的就餐区域。"

Figure 3. BEMBOS Caminos del Inca in Lima, Perú
Figure 4. McDonald's Graz Jakominiplaza Instore

图3. BEMBOS 素食汉堡店，秘鲁，利马
图4. 麦当劳格拉茨Jakominiplaza店

Chapter 2: Site selection

Outdoor seating areas continue to be a large customer draw because those areas provide an inexpensive way to seat more people, attract attention and increase the bottom line.

Particularly in areas with four seasons, people are so happy to be outside when it gets to be springtime,' said Ed Frechette, senior vice president of marketing for Au Bon Pain, in the FastCasual.com article, 'Call of the wild.' 'In the regions where we are — New York, Chicago and Boston — people are thrilled to be able to go outside again and enjoy the fresh air, and that means dining outside.'

室外就餐区相对来说因价格便宜往往能够吸引大量的顾客，从而提高餐厅的收入。

艾德·弗莱切特，Au Bon Pain 快餐集团营销副主席曾在一篇题为《荒野的呼唤》（Call of the wild）的文章中指出："在那些四季分明的地区，人们非常喜欢在春天的时候到户外活动。在纽约、芝加哥和波士顿，人们更乐于外出，呼吸新鲜的空气。在室外就餐，可以说是令人期待的。"

Figure 5. PizzaExpress Plymouth, outdoor dining area
Figure 6. Bembos La Fontana in Perú

图5. PizzaExpress普利茅斯分店,室外就餐区
图6. Bembos素食汉堡拉丰塔纳店，秘鲁

Depending on where the endcap location is, fast-casual operators often have greater access to utilities and other things, such as shopping and office areas. And operators can also take advantage of drive-through accessibility and parking.

'Parking is critical,' said Fox. 'We ideally want to be able to park about 35 cars at our business, but sometimes that depends on cotenants. You could be the only restaurant tenant, but that could change, and then you have competitors competing for parking spaces. If you don't have enough places to put cars, you're not going to have any business.'

Whether a fast casual succeeds as a stand-alone location depends primarily on the concept.

Au Bon Pain currently has outdoor dining at 50 of its 120 locations, and if they could, they would have it at all of their locations, Frechette said.

'The casual segment of the industry has been hurt more than others due to rising costs and the economy,' Bushey said. 'The fast-casual concept offers the casual-dining experience but with more of the speed of a QSR. Dallas, as in many areas of the country, offers the chance to utilize patio dining almost year-round. It has become a very important amenity to offer your customers.'

'There are only so many shopping centres available, and the spaces in those are in demand,' Vais said. 'The trade-off between street-front is you have to spend more in marketing because locations in malls are more compelling for operators. But the trade-off is that your space is oftentimes less expensive. The place where that gets blurred is when you're on a hot street, which also can be difficult to penetrate because of local resistance.'

第二章　选址

弗莱切特提到，目前，Au Bon Pain 的 120 家餐厅内，其中 50 家提供室外餐区。如果可能，他们希望所有的餐厅都能提供室外餐区。

布希提出："随着成本的提高和经济的影响，快餐领域受到严重的冲击。快餐理念旨在提供随意的就餐体验，达拉斯犹如本国的多数地区一样，全年都提供室外就餐服务。这已经成为为顾客提供便利的一种重要方式。"

餐厅的位置往往能够给经营者带来更多的便利，如在商场和办公区附近，就可以充分利用周围的停车条件。

"停车至关重要"，福克斯指出："理想状态上来说，我们需要 35 个停车位，但很多时候取决于合租人。有时你可以是唯一的餐厅经营者，但情况会随时发生改变，需要和别人竞争车位。总之，如果没有足够的车位，那么生意一定不会很好。"

当然，即便餐厅占据单独的位置，是否能够成功更取决于最初的理念。

"购物中心越来越多，而且对于餐厅经营者来说更喜欢选择这里"，维斯指出："和这里相比，临街店面租金相对便宜，但需要花费更多的预算进行市场营销。但是如果在一条繁华的街道上，由于当地各种限制，经营者若想进驻则不是那么容易。"

Chapter 3: Architectural Design

Exterior design

A restaurant's exterior design can say just as much, if not more, to consumers. The exterior of a restaurant is the guest's first interaction with a fast-casual brand, so the façade should appeal to consumers on a visual as well as an emotional level.

Each generation of consumers has a different idea of the type of restaurant they'd like to visit, and with more than 945,000 restaurants in the United States, their options are almost limitless. According to Orlando, Fla.-based Quantified Marketing Group, baby boomers, Generation X and Generation Y (also the millennials) all carry a wide variety of emotional needs.

室外设计

一个餐厅的室外设计可以向顾客展示很多信息。作为顾客与某一快餐品牌的第一次接触，立面设计应从视觉和情感两个方面达到吸引顾客的效果。

每一代的顾客对于餐厅类型的选择都秉承不同的观点。在美国，餐厅的数量达945000之多，因此他们的选择可以说是没有任何限制的。位于弗罗里达州一家名为"Quantified Marketing Group"的餐饮顾问公司曾指出，婴儿潮时期和千福年出生的两代人热衷于根据情感需求选择就餐地点。

Figure 1. Burger King, Goldhill Centre - exterior view
Figure 2. Nando's Dublin - view along the street

图1. 照明设备既能起到点亮空间的要求，同时又可作为装饰元素
图2. Nando's都柏林店——沿街外观

While baby boomers are looking for quiet and soothing restaurant environments and have the money to spend on more upscale locations, those in Generation Y go for more fast-food and quick-service establishments. According to Entrepreneur.com, about 25 percent of Generation Y restaurant visits are to burger franchises, followed by pizzerias at 12 percent.

The exterior of a restaurant must play into what these individual groups are looking for inside as well. The urban and industrial feel of Denver-based Chipotle Mexican Grill appeals to an audience of Generation X and Y; however, that design is in sharp contrast to the softer-focused Panera Bread, which is trying to appeal to a more upscale, sophisticated audience.

Building design

When it comes to exterior building design, earth tones, textures, natural stone and recycled materials are the current trend.

Quick-service chain Taco Bell has incorporated the use of stone in its new building design, and fast-casual Pei Wei Asian Diner has integrated wood and earth tones into its design.

Michelle Bushey, design director and partner with Dallas-based Vision 360 Design, said if a restaurant is going into a strip centre, a lot of the exterior design will be based on the look and feel of the centre. A city or municipality also might have design restrictions in place. For example, the city of Murfreesboro, Tenn., requires a specific percentage of the building to be stone.

婴儿潮时期出生的一代喜欢安静舒适的就餐环境，而且他们有经济能力选择相对高档的餐厅；千福年出生的一代则喜欢快餐或速食餐厅。根据 Entrepreneur.com 调查发现，25% 千福年出生的一代喜欢到汉堡连锁餐厅就餐，而 12% 喜欢到比萨店就餐。

餐厅的室外设计必须体现出其特色，让顾客以此能够了解到餐厅室内环境。位于丹佛的 Chipotle 墨西哥烤肉餐厅运用浓郁的都市和工业风格吸引婴儿潮时期和千福年出生的顾客，然而，Panera 面包店则与其形成鲜明的对比，突出柔和舒适的环境，吸引那些高档消费的顾客群。

建筑设计

涉及到室外建筑设计方面，目前的流行趋势是运用大地色调、天然石材以及回收材料。

Taco Bell 速食连锁餐厅在其新的建筑中采用天然石材，而 Pei Wei 亚洲菜连锁餐厅则运用木材和大地色调。

米歇尔·布希提出，如果餐厅选址在条状购物中心区（沿街道一字排开、有多家商店组成的购物中心）内，那么餐厅的室外设计应该以所在区域的整体形象和环境为基础。另外，每个城市或政府也会有一定的法规限制。例如，位于阿肯色州的莫夫里斯波洛市政府指出，每栋建筑的构造必须运用一定比例的石材。

Chapter 3: Architectural Design

'If you have a Southwest concept, you're going to have to relay the brand message through the exterior and interior design,' Bushey said. 'When people drive by and see the building, it should provide brand recognition.' To keep up with fast-casual players, more quick-service restaurants are updating the look and feel of their interior and exterior designs.

In the QSRWeb.com article, 'Sweet sophistication,' McDonald's spokeswoman Danya Proud said the chain wanted to adapt to the changing lifestyles and tech savvyness of customers, and provide a different experience for in-store guests.

To capture those customers, McDonald's launched a reimaging campaign in 2003.

The company worked with franchisees — who own about 85 percent of the chain's approximately 13,700 locations — and a team of architects, restaurant developers and designers on the colour scheme, design and technological elements of the revamped locations.

The interior and/or exterior design elements of each location vary market to market, but the overall theme remains the same: making McDonald's more relevant and more contemporary to better serve its customers.

'Fast casual is giving the diner the speed of the QSR but with the design of a casual restaurant. So a lot more thought with regard to finishes and materials are put into the interior and exterior of the space,' Bushey said. 'QSR players are changing their look and realizing that if they want to become competitive in this market, they have to step up to the plate.'

Operators are encouraged to work with their design or architectural firm to ensure that each of their desired exterior elements are in place, but an operator should think about what will work and what won't depending on their brand message and niche.

"须通过室内外设计展现品牌标识",布希提出:"当人们开车经过的时候,一看到某幢建筑就可以想到餐厅的品牌。"

为追赶连锁快餐的发展步伐,越来越多的速食餐厅开始提升室内外空间的设计规格。

QSRWeb.com网站上登载的一篇题为《贴心的细节》("Sweet sophistication")的文章中,麦当劳公司发言人丹亚·普罗德提出,麦当劳公司一直在努力追随顾客不断变换的生活方式和技术要求,以便于为他们提供不同的就餐体验。

为吸引顾客,麦当劳公司于2003年启动了一次重新定位品牌形象的活动。

公司与拥有约占品牌85%分店(麦当劳约有13700家分店)的特许经营商合作,其拥有自己专属的建筑师、餐厅发展规划人员和设计师,在色调、设计以及技术方面拥有独特的标准与要求。

Entrances, signage and landscaping

Part of having an inviting exterior includes an accessible entrance and exit.

'You need to provide flow both in and out of the space — you don't want people leaving through the same door they came in if you can avoid it,' Bushey said.

Where the entrance is located should be influenced by how the building is laid out, and it is essential that people know how to enter based on signage, awnings, landscaping, lighting and design.

'Signage is critical, and for us, as a growing brand, it is important for us to get our normal Firehouse sign,' said Don Fox, chief operating officer of Jacksonville, Fla.-based Firehouse Subs. 'It can be frustrating as a

Figure 3. Bembos Aviación — exterior view
Figure 4. Mangiare Spitalfields - shop front

图3. Bembos素食汉堡店圣伯加镇——外观
图4. Mangiare快餐斯皮特菲兹店——店面外观

不同区域的室内外设计元素因市场需求而异，但总体形象保持一致，使得麦当劳公司更具连贯性与现代性，更好地去服务顾客。

"连锁快餐在饮食供应速度上与速食餐厅一样，但其在设计上则突出休闲风格，在室内外空间装饰和材质选择上更加细致用心"，布希解释说："速食餐厅经营者之所以不断地改变自己的形象，因为他们已经意识到若想在这个市场中具有竞争力，必须开始行动。"

鼓励经营者与建筑或设计公司合作，确保关乎室外设计的每个元素都能够恰到好处。当然，经营者自己也必须根据自己的品牌特征和服务范围去思考和选择。

入口、标牌和景观

一个良好的室外环境包括拥有便利的入口和出口。

布希解释说："应该设立不同的入口与出口，以免进出的人群挤在一起。"

入口的位置应该根据建筑的布局确立，而且应该通过标牌、遮阳篷、景观、灯光和设计等结构和元素引导顾客进出。

Firehouse Subs快餐集团运营总监丹·福克斯解释说："对于我们这一正处在发展中的品牌来说，标牌设计至关重要。作为一个经营者，我们在正常使用标牌的时候往往会受到各种限制，这是非常令人沮丧的。一些可能是来自政府方面的规定，另一

Chapter 3: Architectural Design

retailer trying to grow and face some of the restrictions that we do on using our usual signage. Some may be government imposed, but then you have the developer who may impose it to keep with the feel of the centre they're designing.'

Fox said that typically, customers want an easily identified retail establishment rather than an unbranded one.

'When you see the golden arches, you know it is McDonald's,' he said. 'You just have to weigh the benefits of the location versus using your signs. If it's an existing centre where all of the other tenants have conformed, it will be more difficult for a developer to change rules. The greatest chance of success is if you've gone in from the beginning.'

方面可能来自开发商要求与他们的整体环境保持一致。"

福克斯还提到，顾客希望看到的是一个易于辨识的品牌标识，而不是毫无自身特色的。

"当看到一个金黄色拱形标志，你一眼便知这是麦当劳"，福克斯补充说："必须去衡量餐厅选址和品牌标识之间的利害关系。如果在一个既定的商圈中，其他的经营者全部遵循现有的风格，那么很难让开发商再去改变。在这种情况下，如果想要获得成功，那么从开始就应该介入其中。"

Figure 5. McDonald's Haus im Ennstal - the building seems emerged in the surroundings
Figure 6. Logo of McDonald
Figure 7. Bembos La Fontana - Logo on the building

图5. 麦当劳豪斯恩斯河谷地区店——建筑似乎与景观融合在一起
图6. 麦当劳标识
图7. Bembos素食汉堡拉丰塔纳店——建筑外观上醒目的标识

第三章　建筑设计

In the book 'Your New Restaurant,' author Vincent Mischitelli suggests that an effective sign is one made by experts. The sign also should be clear and simple, and let customers know what they're coming into the restaurant for.

Landscaping elements are another big draw into eating establishments, as they can bring in new customers and play into the overall look and feel of the restaurant.

'If you're doing a ground-up project, most of the time it is wise to hire a landscape designer or architect. If you go into a strip or lifestyle centre, the landscaping has already been dictated since the developer provides it for the site,' Bushey said.

Shrubs have a one-time cost but can last for several years, while flowers add a nice touch of colour in the spring and summer months.

'A well-designed landscape can enhance the appearance of a restaurant on a year-round basis,' Mischitelli said in his book. 'A lined parking lot can add a feeling of organization and professionalism to an establishment.'

Definitions
Baby boomers: Born between 1946 and 1964, this group makes up the largest segment of the U.S. population.
Generation X: Born between 1965 and 1977, this group is known for having strong family values.
Generation Y: Born between 1980 and 2000, this generation is ethnically diverse and three times the size of Generation X.

名词解释
婴儿潮时期出生人口：1946年至1964年在美国出生的人群，他们构成了美国人口的大部分。
X一代：1965年至1977年在美国出生的人群，他们通常具有很强的家庭观念。
Y一代：1980年至2000年在美国出生的人群，民族多样化，人口数量约为X一代的三倍。

文森特·米斯奇特里曾在《你的新餐厅》（Your New Restaurant）一书中指出，一个最佳的标识应该出自专家之手，简洁清晰，让顾客明了他们来到餐厅即将会带来的体验。

景观元素在餐饮空间中格外重要，巧妙的设计可以帮助吸引新的顾客，同时为餐厅的整体形象增添魅力。

布希曾这样解释："如果是一个整体设计的项目，那么明智的选择是雇用一个景观设计师；如果是进驻条状购物中心或者休闲生活中心，那么景观环境已由开发商负责打造。"

树丛是一次性成本投资，但却可以持续多年，而花丛则可以在春季和夏季增添多样的色彩。

米斯奇特里也曾在书中提到，良好的景观元素可以在全年为餐厅带来全新的面貌，画线停车场可以增强餐厅的组织性和专业性。

Chapter 4: Interior design

Interior design

When Robert LaGore was determining the look and feel of his fast-casual restaurant, Marion,Ill.-based Burgers-N-Cream, he started with ideas from more than 10 existing concepts. One of the brands he closely studied was that of Denver-based Chipotle Mexican Grill.

'I think there are a percentage of people that go into Chipotle that are drawn back by the feel,' LaGore said. 'I'm a big believer there's not a person out there that the design is going to draw them into the restaurant, but I'm a big believer it's going to draw them back.'

LaGore wanted to ensure he had the right look and feel for Burgers-N-Cream, a hamburger and ice cream chain designed around the Americana visions of apple pie, ice cream and grandma.

The restaurant has a professional feel during the lunch daypart and transforms into a casual, family-friendly atmosphere when the evening rush hits.

'Our design has been a little different because we try to promote cooked-to-order freshness, which is what differentiates us,' LaGore said. 'Our whole design is to show a fresh, clean product.'

Before LaGore could get into the design phase, he needed to know his demographic and what his concept stood for. He then needed a design strategy that complemented the brand and met his customers' expectations.

'The brand has to work with the actual design. It has to look like what they're selling. It has to be cohesive. I think, just like Robert said, your design is not going to bring people in the door, but it's going to get them back,' said Tami Stallings, an interior designer for Landmark Design & Engineering Inc., who worked with LaGore on the design of Burgers-N-Cream.

室内设计

罗伯特·拉格尔在为其经营的位于马里恩市的汉堡与冰激凌（Burgers-N-Cream）连锁快餐厅设计外观与风格时，参考了10余家品牌餐厅的设计理念，其中深入研究了位于丹佛的Chipotle墨西哥烤肉餐厅的设计。

"我认为，一定数量来过Chipotle就餐的顾客会被其风格所吸引，并再次光顾"，拉格尔指出："我非常相信餐厅想要吸引的顾客一定会光临这里，而且我更相信，他们之后一定会再次光临。"

拉格尔想要确保为自己的品牌餐厅——一家围绕美式苹果派、冰激凌和老奶奶为主要理念的连锁餐厅，设计合适的外观与风格。

餐厅在午餐供应时间突出专业特色，而在夜晚时分则转换成随意友好的家庭式氛围。

"我们的设计少许有些与众不同，因为我们努力推崇量身定做的食物"，拉格尔解释说："我们的设计就是围绕提供新鲜、干净产品的理念而开展的。"拉格尔在进入设计阶段之前，充分了解该地区的人口因素以及明确其经营理念。之后，他制定与品牌特色相辅相成的设计计划，以满足顾客的需求。

"品牌特色应与空间设计相互统一，应展示出餐厅产品的特色，更应该连续一致。正如罗伯特所说，我认为一个成功的设计不仅将顾客领进门，更重要的是让他们时常光顾"，塔米·斯塔灵思（地标设计与工程公司室内设计师，曾与拉格尔合作）如是说。

第四章 室内设计

In addition to ensure that the interior design fit with the brand, Stallings was charged with ensuring an efficient flow of both back- and front-of-house operations.

此外,为确保室内设计与品牌特色协调一致,斯塔灵思在设计中确保了餐厅前后空间的流畅性。

Figure 1. BEMBOS Caminos del Inca in Lima, Perú

图1. BEMBOS 素食汉堡店,秘鲁,利马

Chapter 4: Interior design

Figure 2/3/4. Mangiare in London - the straight layout showing fluidity of the space

图2/3/4：Mangiare快餐伦敦店——直线型的格局凸显空间流畅性

Design for function: front and back of house

When it comes to any restaurant operation, an efficient, well-designed kitchen is essential.

'In the kitchen, the more efficient the design, the quicker you can provide products and services, which results in more turns on covers in the front of house,' said Michelle Bushey, creative director and partner with Vision 360 Design based in Dallas.

Bushey said it's important to have a kitchen designer who can help lay out the kitchen so it will flow based upon the menu and concept. 'Certain menus are going to require specific types of equipment,' she said.

In addition, the kitchen and the wait stations should be blocked out at the beginning of the design-development process.

Other items to take into consideration when designing the kitchen include room for storage and expansion, including equipment upgrades.
LaGore's kitchen design is somewhat different from other fast-casual operations.

'One of the things we do that's unique is we cut our fries up on the line,' he said. 'Any minute one of the cooks is not busy, they're cutting potatoes. So when you pick up the food, you're seeing fresh-cut potatoes right on the line.'

The kitchen also has an open layout so customers can see their burgers and other items as they are being prepared.

'We're showing everybody everything is fresh,' he said. 'On back side, we're 95 percent comfortable with the flow in our kitchen.'

功能设计——前店与后屋

对于任何一个餐厅来说，一个有效而精心设计的厨房都是至关重要的。

"在厨房中，越是有效的设计，越能提高产品供应的速度，同时就能保证前店能够接待更多的顾客"，米歇尔·布希解释说。

布希还指出，厨房设计师可以帮助规划厨房布局，确保其能够以餐厅菜单和设计理念为基础而正常运转。"特定的菜肴需要特殊类型的设备，因此需要经过仔细规划"，她补充说。

此外，厨房和等候区在设计之初就应该被分隔出来。在厨房设计中还应考虑的其他因素包括存储空间、扩建空间以及设备升级等。

拉格尔在厨房设计上突出了与其他快餐厅的区别。"我们最为与众不同的地方就是薯条现炸现卖"，他解释说："大厨们闲下来的时候，他们就会切土豆，所以当你来拿食物的时候，会发现还未经煎炸的土豆条。"

此外，厨房的开放式格局让顾客可以观赏到汉堡或者其他食物的制作过程。

"我们乐于向每一位顾客展示我们这里的所有食物都是新鲜的，而且我们的厨房舒适度和效率高达95%"，拉格尔解释说。

Chapter 4: Interior design

When it comes to front-of-house design, the idea is not to crowd customers but enable them to get their food items without hassle, Bushey said. 'You don't want dead ends.'

At Burgers-N-Cream, every area is showcased with a neon sign, drawing customers' attention to specific regions of the restaurant where they can pick up their drinks, ice cream or food items.

One sign highlights the premium ice cream service area, while another draws attention to the beverage station. And a canopy across the top of the kitchen area reads 'Cook to Order,' which also conveys the restaurant's message of freshness.

Design elements

Every restaurant environment will be different, and every fast-casual restaurant will fight to differentiate itself from other brands. In the design

关于餐厅前店的设计，首先不要让顾客感到拥挤，然后让他们能够顺利拿到自己的食物。"谁也不想走进死胡同"，布希解释说。

在汉堡与冰激凌餐厅（Burgers-N-Cream）内，每一个区域都采用霓虹灯标识，指引顾客在不同的区域挑选食物。

其中一个标识指向冰激凌区，而另一个则指向饮料区。横跨厨房上方的标识上写着"Cook to Order"（量身定做），突出了餐厅的品牌特色。

设计元素

每一家餐厅的环境都是不同的，每一家连锁快餐厅都努力使自己与其他品牌不同。在设计领域中，有

Figure 5. Nando's Dublin in Mary Street - patterns on the wall adding vividness to the dining area

图5. Nando's都柏林店，玛丽街——墙面上的装饰图案增添了空间趣味性

world, the bold, and sometimes architecture, menu and graphics all need to work together and relate to one another to produce a successful brand 'package.'

'There are times when they don't, and although most customers can't necessarily pick out what isn't working within the package, they can definitely sense it,' Bushey said. 'This is where a design professional that specializes in restaurant and hospitality design really makes a difference.' Understanding the dynamics of the kitchen design and efficiency, knowing customer and staff traffic patterns, seating options, obnoxious, colours of the 1980s and 1990s paved the way to the use of the earth tones and softer tones found in many restaurants today.

'Depending on the concept, we're starting to see more designer-oriented finishes coming into the space,' Bushey said. 'But you also have to bring in items that are functional and cost sensitive for the brand. I can't put a $300 chair in a fast casual, but I can see a $75 chair. Everyone loves a wood or concrete floor, but I have vinyl that provides the same look at a lower cost that is easier to clean and requires less maintenance.'

Obtaining and maintaining cohesiveness within a brand is a key to design strategy. The logo, interior design, etc., will optimize your design, increase your productivity in the kitchen, maximize your seating capacity and invariably increase your profits.

Starbucks made a big impact on the design and restaurant industry when it launched its first coffeehouse in downtown Seattle in 1971.

'Everybody wanted to capture that environment in their space,' Bushey said.

Today, many fast casuals are still trying to appeal to harried consumers by offering laid-back surroundings where guests feel they can linger. Chairs play a large role in that design element, with fabric overtaking vinyl.

第四章　室内设计

时大胆的精神、空间、菜单、平面需要融合在一起，相互联系，打造一个成功的"品牌包。"

"有时，这些元素并不能很好地统一起来，尽管大多数顾客不能指出到底是哪里不和谐，但是他们能确切地感受到这种不和谐的存在"，布希曾解释说："这就是餐饮领域专业设计师擅长解决的问题。"了解厨房设计的动态性和效率性、知晓顾客和员工在餐厅内的行走路线、熟悉餐厅装饰中色彩选择的趋势，这些对于空间设计至关重要。

"现如今，我们发现越来越多设计师定制的装饰元素开始在餐厅空间中流行起来"，布希解释说："其实经营者可以自己选择那些物美价廉但适合品牌特色的装饰品。我不可能将价值300美元的椅子放到快餐厅中，但是我可以选择75美元的。人人都喜欢木质或者水泥地面，但是我运用树脂材质可以设计出同样的外观，不仅造价低而且易清洗、易维护。"取得并保持连贯性对于设计来说至关重要。标识以及室内装饰可以优化整体设计、提高厨房生产率、增加顾客就座率，从而提高营业利润。

星巴克1971年在西雅图开设的第一家咖啡店对于整个餐饮业带来了巨大的影响。

"每个经营者都想营造出那样的环境与氛围"，布希指出。

如今，很多快餐厅一直在努力吸引那些奔忙的顾客，通过营造悠闲的环境使他们驻足停留。椅子成为了设计元素中的重要部分，而且织物取代了皮革。

Chapter 4: Interior design

'We're seeing higher-end fabrics, new finish technologies; we're seeing more modern and comfortable furnishings and a lot more booths,' Bushey said.

"我们现在有高端的织物饰品、全新的装饰技术，我们有更加现代和舒适的装饰结构，我们还有幽静私密的雅座"，布希解释说。

对于很多餐厅经营者来说，雅座和椅子相结合的方式似乎是希望将顾客留下来。他们还乐于寻找一些简单的椅子装饰物。

Figure 6. McDonald Concept Racine in Villefranche-de-Lauragais (31), France

图6. 麦当劳概念店，法国，洛拉盖自由城——简约风格的桌椅营造出舒适的就餐氛围

第四章　室内设计

For many restaurant operators, a combination of booths and chairs seems to hold something for everyone. Operators also are looking at a variety of chair finishes and items with very clean lines.

'What those customers need to look at is the quality and construction of the product,' said Bill Bongaerts, president of Winston-Salem, N.C.based Beaufurn LLC. Beaufurn provides commercial furniture in the restaurant and bar space.

'Typically, people in that field look for the cheapest product. If you give up on budget, give up on quality of product as well,' Bongaerts said. 'Even in fast-food environments, those people are going upscale. Five or six years ago, that was unheard of.'

A big design push also is taking place with regard to lounge seating. Bongaerts said the push stems from operators trying to create a more relaxed and inviting environment, one where people can sit comfortably for lunch or linger over dinner.

'People can come in and see something that's comfortable as opposed to in and out. It's not just the seating; it's the quality of food as well. Everything is changing,' he said.

Sandra Saft, president and founder of Orlando, Fla.-based Windows Interior, said restaurants are moving away from the use of mini blinds, which get dirty easily and can make the entire restaurant seem unclean. 'But in the last 15 years or so, the industry has moved away from mini blinds into solar-shading ideas,' Saft said.

Solar shades keep out heat and the sun's glare, without blocking natural light. 'The concept is so simple, yet it has so many positive aspects to it,' Saft said.

"顾客最关注的就是产品的质量和构造",比尔·班加尔特(Beaufurn LLC 家具公司负责人)强调说。"通常情况下,经营者喜欢购买最便宜的物品。当然,你在价格上让步,那么在质量上也不能要求太高",班加尔特解释说:"即使是在快餐环境中,顾客们也会喜欢高档的氛围。当然,这在5、6年前是闻所未闻的。"

一个较大的推动元素即为休闲椅的运用。班加尔特曾提出,这一推动力源于经营者试图打造一个更加随意舒适的环境,让顾客可以舒服地享用午餐或晚餐。

"顾客有时会进来看一看环境是否舒适,然后决定离开或者留下来。当然这不仅是指座椅,食物的质量同样重要。一切事情都在变化",他解释说。

桑德拉·萨夫特,奥兰多 Windows Interior(室内窗)公司主席和创始人,指出餐厅逐渐开始摒弃小型百叶窗的使用,因为它们特别容易变脏并使得整个餐厅看起来不整洁。

"从15年前开始,遮阳屏逐渐取代了小型百叶窗",萨夫特解释说。

遮阳屏可以阻挡热量和太阳强光反射,但却不会阻止自然光线的进入。"这一理念非常简单,但是却能带来很多有利的影响",萨夫特补充说。

Chapter 4: Interior design

Chipotle Mexican Grill and Panera Bread are two examples of fast casuals using solar shades in restaurants.

The newer shades are polyester-coated woven materials that also are biodegradable. Further, they can be washed or hosed off without causing damage to the shade.

'It's a fat shade screen instead of a curtain, so it's cleaner and allows the light to come in with whatever density the selection may be,' Saft said. 'You can also still use a valance or tieback curtains, which give it a design element. Also, it's such a nice, clean look all by itself.'

While the solar shades control some of the light, more should be done to make guests feel comfortable in the environment.

Chipotle 墨西哥烤肉餐厅和 Panera 面包店是两家利用遮阳结构的典型。

较新的涂层织物材料遮阳屏可降解，而且可以水洗或者刷洗。

"这是宽大的遮阳屏而非帘子结构，可以清晰地过滤出哪种密度的光线可以进入"，萨夫特解释说："当然，也可以使用帷幔或挂钩窗帘。这些都是设计元素。"

遮阳结构虽然可以控制光线，但仍需更多元素以便于为顾客营造舒适的就餐环境。

Figure 7. Nando's Dublin in Mary Street - lighting fixtures dangling from the ceiling brightening the entire space
Figure 8. Sosushi Rho - The spot light punctuates the rhythm of the ceiling illuminating a cascade of 1,000 origami hanging in the air

图7. Nando's都柏林店，玛丽街——灯饰从天花上垂落下来，点亮了整个餐厅性
图8. Sosushi寿司店罗镇分店——聚光灯打破了天花的节奏，纸鹤 在灯光的照耀下更加引人注目

Lights, music, acoustics

In addition to adding comfortable seating options, lighting, music and acoustics play a significant role in interior design.

'It is getting better, but this was almost always the most overlooked, and one of the first things cut from the design budget, and is probably one of the most important aspects of design,' Bushey said. 'Lighting can create intimacy or energy depending on how you use it, and it can make your product look better as well. In many ways it brings the concept, architecture and design together to create that emotion or evoke the feeling the owner wishes to convey to the customer.'

Lighting, coupled with the right acoustics, also can enhance the customers' entire dining experience. When diners sit at Rubio's Mexican Grill, they are transported to another time and place through the sounds of the surf, seagulls and crashing waves.

'When people walk in there, they feel like they're in Baja,' said marketing expert Linda Duke. 'The specifically chosen songs and sounds give guests the feeling that they are sipping their Corona and eating a fish taco with their toes in the sand and the ocean rolling in.'

While it doesn't provide 'beach music,' Chipotle Mexican Grill pipes in music to each of its more than 670 units. The music is designed to create a relaxed environment and fit into the urban, industrial design of its units. A person can sit in any number of fast-casual restaurants and notice the sound of music from above, but no acoustical element — unless it's live music — should get in the way of customers' conversations.

'Everyone loves open ceilings and concrete floors, but you need something to absorb the sound,' Bushey said. 'Soft seating such as booths and banquettes, fabrics, acoustical panels, dropped ceilings with tiles, vinyl flooring, etc., will all help to absorb and dampen sound. No one likes to have to yell across the table at each other while trying to enjoy a meal.'

灯光、音乐与音响

除了舒适的座椅之外，灯光、音乐以及音响效果在室内设计中同样起着重要的作用。

"如今这些情况已经变得越来越好，但还是经常被忽略。很多情况下，最先从中减少预算的部分往往是设计中最为重要的部分"，布希解释说："灯光可以帮助营造亲切的环境，可以产生能量，还可以让食物更加突出，当然取决于如何利用。灯光可以通过多种方式将品牌理念、建筑与设计结合在一起，打造出一种经营者最希望传递给顾客的氛围。"

灯光配以合适的音响系统可以丰富顾客的就餐体现。比如，在 Rubio's 墨西哥烤肉餐厅就餐时，伴随着浪花和海鸥的声音，仿佛到了另一个时空。

"走在 Rubio's 餐厅内，如同置身于下加州"，营销专家琳达·杜克解释说："顾客喝着科罗娜啤酒，品尝美味的鱼香夹饼，听着餐厅专门选择的歌曲和声音，仿佛在海滩上野餐。"

与海滨音乐相比，Chipotle 墨西哥烤肉餐厅则选择了更适合其都市和工业风格的音乐，并在其 670 多家分店中播放，营造了一个轻松随意的就餐环境。

顾客可以发现在任何一家快餐厅中，音乐都不会影响到顾客之间的交谈，除非是现场演唱。

"每个人都喜欢开阔的天花和水泥地面，但是我们需要一些结构来吸收声音"，布希指出说："软座区如雅座或者长椅、织物、吸音板、瓷砖吊顶、树脂地面等都可以帮助吸收或减弱噪音。没有人喜欢在吃饭的时候隔着桌子大吼说话。"

Chapter 5: Considering the queue

Consider the queue

In any fast-casual environment, it's important that customers have an easy and clutter-free order experience. It's also important that the collateral surrounding the point-of-sale tells your brand's story.

Whether a customer decides to wait or leave a busy restaurant hinges on the time of day and the quality of the food.

'It depends on what customers are willing to deal with,' Bushey said. 'If it's 12:30 p.m. and a hot place for lunch, people tend to have more patience since they know what to expect. Some people might be turned off because they don't like waiting in line. If you make the order process uncomfortable and customers are on top of each other all the time, it's definitely going to be an issue.'

When it comes to your queue, Bushey offers four recommendations operators can use to make the process as carefree and inviting as possible.

1. Put the location of the queue in an area that eliminates confusion and drives customers to the POS.

2. Provide an order experience that tells your brand's story.

3. Place POP marketing and the menu board within the queue to drive sales and speed up the order process.

4. Make sure flow/traffic patterns allow maneuverability and ease congestion for both customers and staff.

The fresh-Mex segment provides perhaps the best example of moving customers through the line during their order experience. Chipotle, Qdoba and Moe's Southwest Grill each provides an order area for customers

排队因素

对于任何一个快餐厅来说，为顾客带来简捷便利的点餐体验都是非常重要的，同时收银台周围的装饰布置最好能够述说关于品牌的故事。

顾客选择在餐厅中等待或者是离开，在一定程度上取决于时间段和食物的质量。

"有时候取决于顾客自己的想法"，布希解释说："如果是在中午12:30，而且是一个午餐比较知名的地方，那么顾客也许会有耐心等待。当然，也会有一些人不喜欢排队而选择离开。如果点餐过程很不顺利，而且队列混乱，这更是一个大问题。"

提到排队问题，布希提出四个建议供经营者参考，以便于使点餐过程变得畅快舒适。

1. 排队区标识指示清晰，指引顾客走向收银台，避免让顾客感到迷惑；

2. 给顾客提供一种可以讲述品牌故事的排队体验；

3. 将餐厅宣传册和菜单放到队列区内，以加快点餐过程并促进消费；

4. 确保队列流动的方式具有可调控性，避免顾客和餐厅人员的拥挤。

墨西哥快餐厅在顾客排队点餐的流程设计中提供了最好的例子。例如，Chipotle，Qdoba以及Moe's墨西哥餐厅都设定专门的点餐区，顾客可

第五章　排队因素

Figure 1. McDonald Concept Racine in Villefranche-de-Lauragais (31), France - enough space for customers waiting in the queue when ordering the food

图1. 麦当劳概念店，法国，洛拉盖自由城——餐厅中预留了足够的空间供顾客排队等候点餐

Chapter 5: Considering the queue

that let them pick the ingredients to be used on their menu item as they move through the line. Other fast-casual chains, such as Penn Station East Coast Subs and Firehouse Subs, offer the type of order experience where customers order at one area then move down the line to the POS.

The design shift gives employees the ability to face the customer as their sandwich is being prepared, significantly reducing the order-taking process, said Penn Station president Craig Dunaway.

'We move you down the line instantly,' Dunaway said. 'A process that used to take six to eight minutes, we've really moved that up between four and six minutes. It has allowed stores to grow sales because it's expediting the order process. They turn tables faster, and it just snowballs in a positive way.'

Telling your brand's story

Just as important as driving customers to the POS is educating them about your brand's story.

When Chris Dahlander worked in marketing for Romano's Macaroni Grill, he used danglers, table toppers and anything else he could think of to tell the brand's story. But now that he owns Snappy Salads, a fast-casual restaurant in Dallas, Dahlander has fine tuned his approach.

His restaurant features table merchandisers — simple, brown recycled Kraft paper with a printed letter from Dahlander to guests, personally signed by him. Once a month, he changes the merchandisers to discuss a different part of his story: the biodegradable containers he uses, the charities he supports, the evolving soup selection.

'I try to touch every table, but at the same time I know people have busy lives so I can't tell them a 15-minute story about my life and why I'm doing what I'm doing. Table merchandisers help a lot,' he said in the FastCasual.com article, 'Deliver your message with POP.'

以在排队过程中选择他们想要购买的食物。其他快餐品牌，如 Penn Station East Coast Subs 则带来另一种体验——顾客在一个区域点餐之后，排队去收银台付款。

Penn Station 连锁集团主席克雷格·邓纳维指出，设计方式的改变让餐厅员工在食物制作的过程中可以直接面对顾客，从而简化订餐流程。

"这种方式大大缩短了点餐时间，由原来的6至8分钟，缩短到4至6分钟。如此一来，就可以增加食物的出售量，从而增加营业额。"邓纳维解释说。

述说品牌自己的故事

指引顾客走向收银台的过程中向他们讲述品牌的故事，是一种重要而有效的营销方式。

克里斯·达兰德在为 Romano's 通心粉餐厅做营销计划时运用了吊饰、桌垫以及任何他能够想到的、可以展示品牌特色的物品。现如今，达兰德拥有了属于自己的连锁快餐厅，Snappy Salads，他换用了另一种方式讲述品牌故事。

他的餐厅以桌子上的特色而著称，简单而可循环利用的牛皮纸上印着达兰德写给顾客的话语，并有他的亲笔签名。每隔一月，他就变换上面的内容，讲述其他的故事，如餐厅中使用的可降解餐具、他举办的慈善活动以及更新的汤类食品菜单等。

"我尝试在每张餐桌上写上关于我的故事，但是故事不会超过15分钟，因为顾客很忙，他们不会

第五章　排队因素

The role of point-of-purchase materials is to tell a story, not sell the product, said international sales and marketing expert Bob Phibbs.

A brand's story can be told through POP materials near the POS, through the menu board and through employees, who will always be the best profit-drivers in any restaurant environment.

'It's all about the order experience, because after all, the object is to get people in line and back through your door,' Bushey said. 'If you've done your due diligence, have good staff and a quality product, you should be able to hit those marks.'

抽出更多的时间去关注这些",邓纳维曾在一篇在FastCasual.com 网站上发表的文章中补充说:"这种方式带来了很多帮助,其实以一种广告的方式传递了品牌特色。"

国际营销专家鲍勃·菲比斯指出,售卖场所广告(point-of-purchase)材料并不是在售卖产品,其实是在讲述一个故事。

一个品牌的故事可以收银台附近的广告(POP)材料讲述,也可以通过菜单和餐厅的员工讲述,他们往往成为餐厅经营中利润的带动者。

"关于点餐经历,最重要的是让顾客感觉顺利,然后能够再次光临餐厅",布希解释说:"如果你足够勤奋,拥有高素质的员工,提供高品质的食物,那么一定会成功。"

Figure 2. Sosushi Train Turin, cherry tree pattern highlighting the whole inteiror

图2. 寿司店都灵火车站分店,樱花树图案成为整个室内空间的亮点

McDonald's
麦当劳

In 2006, McDonald's introduced its 'Forever Young' brand by redesigning all of its restaurants. McDonald's has invested $1 billion to redesign nearly all of the 14,000 restaurants by 2015.

The design includes wooden tables, faux-leather chairs, and muted colours; the red was muted to terra cotta, the yellow was turned golden for a more 'sunny' look, and olive and sage green were also added. To warm up its look, the restaurants have less plastic and more brick and wood, with modern hanging lights to produce a softer glow. Other upgrades include double drive-thrus, flat roofs instead of the angled red roofs, and replacing fiber glass with wood. Also, instead of the familiar golden arches, the restaurants now feature 'semi-swooshes' (half of a golden arch).

2006年，麦当劳集团开始贯彻"永远年轻"（Forever Young）的品牌承诺和精神，计划投资10亿美元于2015年前完成对全球约14000家餐厅的改造。

设计元素包括木桌子、人造皮革座椅和中性色调。赤褐色代替大红色，黄色转变成金色以打造更加"阳光"的外观，橄榄绿和灰绿色被引入进来。为营造更加温馨的氛围，塑料材质多被砖石和木材取代，现代化的灯饰被大量运用。其他元素包括双车道、平屋顶等。此外，麦当劳广为熟知的金色弧形标志也被半弧图案取代。

McDonald's Concept Racine
麦当劳概念店

Completion date: April 2011
Location: Villefranche-de-Lauragais (31), France
Designer: Studio Patrick Norguet
Photographer: Renaud Callebaut
Area: 240 m²
Promoter (client): McDonald's France
Lighting: Gerard Foucault

完成时间：2011.4
地点：法国，洛拉盖自由城
设计：帕特里克·诺尔盖设计工作室
摄影：雷诺·卡勒伯特
面积：240平方米
客户：麦当劳法国公司
灯光设计：杰拉德·傅科

McDonald's has put Patrick Norguet in charge of designing the new architectural identity for its restaurants in France. A project which is exciting in terms of its scope as well as in its technical and sociological constraints since it concerned McDonald's returning to its founding myth: familial fast food. If the brand was originally founded on the family, its image has little by little slid towards a more urban and adolescent tone. A return therefore to McDo's DNA with this new interior design that Patrick Norguet, literally and figuratively, matches with getting back to roots.

The plant metaphor, with its branching development, this root common to the brand and to the family, is transformed here into an architecture which is transversal and expansive: birch plywood takes root and branches out in the restaurant in order to create areas, functions and moods for different social requirements without compartmentalizing.

This organic and functional furniture/architecture offers several possibilities, several eating choices from eating standing up for lone teenagers, alcoves providing privacy to family table service, a small revolution at McDonald's with digital control terminals integrated into the base and distributed throughout the restaurant. Henceforth, a mother can settle with her offspring at a table, order from a nearby terminal and wait for the meals to be brought to the table.

Patrick Norguet's design, which as always hits the spot, uses contemporary white which he counterbalances with fun colours without falling for 'toy' conventions like for example the storage elements with the painted metal boxes included in the base template. The luminous ambiance and the quality of the acoustics are exceptionally meticulous and offer customers comfort which is rare today, whilst the quest for a certain radical nature is revealed through the choice of materials (plywood, sheet metal, concrete, etc.), tested in conditions of heavy passage to respond to the constraints of such a popular restaurant.

The designer is using his 'Still' metal chair for Lapalma for the seats with a new high stool version specially designed for the occasion. The ceramic floor also designed by Patrick Norguet for Lea Ceramica immediately lends a distinctive tone to the venue. These huge, ultra-slim 2-metre slabs break with usual visual conventions: warm and graphic without being carpet, they change our habits in terms of flooring to create a brand new typology.

Piloted at the start of the year in the Villefranche-de-Lauragais restaurant 40 km from Toulouse, the concept was immediately appealing and spoke volumes. 6 restaurants are currently in the pipeline throughout France.

麦当劳快餐集团委托帕特里克·诺尔盖负责为法国地区的餐厅打造全新的建筑形象。无论是从范围还是技术及社会的制约等方面来讲，这一任务都将是令人期待的。麦当劳的形象逐渐变得都市化并趋向青少年人群，因此这一项目设计的主旨即为找回麦当劳创立之初的根本理念——"家庭式快餐。"帕特里克·诺尔盖成功的实现了全新的空间设计与根本理念的完美融合。

设计师将这里比喻成一棵大树，"根部"便是麦当劳特有的品牌理念，"枝叶"象征着不断延伸与发展。多层桦木板被用作设计元素，在这里生根、发芽，将餐厅在功能和氛围上划分成不同的小空间，与传统的空间分割方法相比更加灵活。

空间的巧妙划分为顾客营造了多样的就餐区域，如供单独就餐的年轻人吧凳区、供全家就餐的私密小空间等。此外，终端数字控制器在餐厅中的广泛应用为顾客点餐带来了更多的方便。比如说，带孩子的年轻妈妈可以通过餐桌附近的控制器直接点餐，然后只需等待食物被送来就可以了。帕特里克·诺尔盖的设计总是能够恰到好处。他采用现代风格的白色与趣味性十足的色调搭配，并能够在两者之间实现平衡。例如，白色储藏柜内间或存在的喷漆金属盒子。灯光氛围以及音响效果的打造格外注重细节，为顾客带来一种鲜有的舒适感。然而材质的选择则完全打破常规，多层板、金属板及混凝土的运用别具一格。

设计师自己打造的金属凳子被新添了一个靠背，与整体空间氛围契合，而设计师专门设计的地面则为空间带来了些许的与众不同。地面由细小的板块拼接而成，打破视觉常规以及传统的地面铺设方式，营造了一种全新的特色。

这一项目于2011年年初竣工，其理念受到一致的赞同。随后，法国地区共有6家麦当劳餐厅开始使用这一理念。

1. Touches of bright colours making the space more vivid
2. Digital control terminals integrated with dining area
3-4. Dining areas of various styles
5-6. Painted metal boxes included in the base template

1、几抹亮丽的色彩使得空间更加生动
2、与就餐区相连的点餐控制终端
3、4.不同风格的就餐区
5、6.彩色喷漆金属格子别具特色

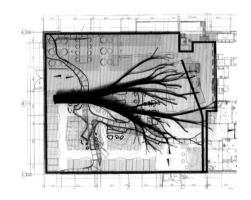

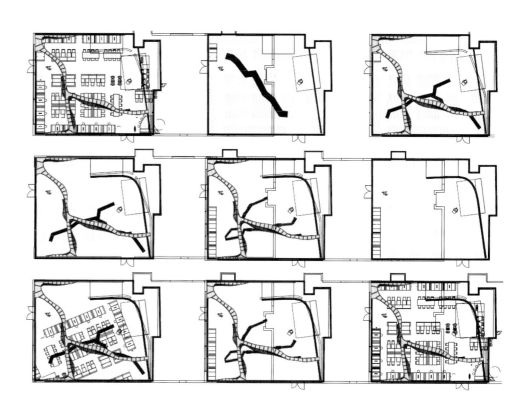

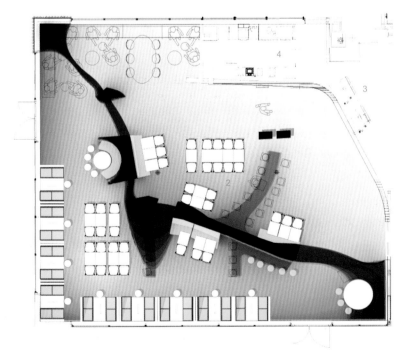

1. Entrance
2. Dining area
3. Order bar
4. Kitchen

1. 入口
2. 就餐区
3. 点餐台
4. 厨房

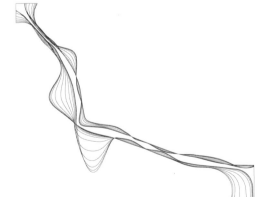

Floor plan 平面图

McDonald's Graz Jakominiplaza Instore

麦当劳格拉茨Jakominiplaza店

Completion date: April 2010
Location: Styria, Graz, Austria
Designer: Viereck Architects Ltd.
Photographer: Foto Schmickl
Site area: 610,00m²
Construction area: 603,50m2(97 seats restaurant+36 seats McCafé lounge)
Client: McDonald's Franchise GmbH

完成时间：2010.4
地点：奥地利，格拉茨，斯蒂里亚
设计：Viereck建筑有限公司
摄影：Schmickl 摄影
占地面积：61000平方米
建筑面积：60350平方米（餐厅97个座位，咖啡厅36个座位）
客户：麦当劳连锁有限公司

In the centre of Graz, in the historical part of Herrengasse and Jakominiplatz, this instore-restaurant is located. It was the first McDonald's restaurant in Styria around twenty years ago.

Once the restaurant is now showing its age, and of course the quality of the interior design has changed; it was decided in 2009 to clean up the object completely and revitalize. To be prepared for the future, the building was in the basement, ground floor and first floor fully hollowed out and stripped down to the load-bearing walls.

In a relatively short construction period, the old building structure was brought up to date and changed into a restaurant with modern amenities and quality interiors. On the ground floor a new McCafe was situated with a large comfortable lounge. The colour concept tries with different shades of brown to create a harmonious, peaceful entrance and the design concept got also a different seating layout, starting with lounge seating and stand-ups for busy guests.

Upstairs a large guest area with about 100 seats were created, which is equipped same as on the ground floor with various seating qualities. The colour scheme was carried out in largely muted, earthy colours and creates a nice, soothing ambiance. High-quality decorative materials in wood design, as well as graphic printouts on the walls invite the guest to dwell on the leather-upholstered benches. The existing openings of the historic building have been retained and with glass panels endorsed in order to provide an overview from the lounge to the dynamic space.

1. Identifiable signage of McDonald's
2. Facade boasting combination of classic style and modern feeling
3. Order counter
4-5. Dining area featuring chairs of various colours
6-7. Different shades of brown to create a harmonious, peaceful atmosphere
8. Staircase leading upstairs
9-12. Dining areas of different styles

1. 麦当劳特色标识
2. 外观凸显传统特色和现代风格的融合
3. 点餐台
4、5. 就餐区内色彩丰富的座椅别具特色
6、7. 基于棕色调的多种色彩的运用营造了一种和谐而平和的氛围
8. 通往上层的楼梯
9~12. 不同风格的就餐区

项目选址在格拉茨绅士街（Herrengasse）和Jakominiplatz广场历史中心区，是斯蒂里亚地区20年前第一家麦当劳餐厅。

鉴于餐厅较长的历史以及室内设计的质量，2009年开始决定进行改造。为适应未来的发展规划，整幢建筑，包括地下室、一层及二层，被彻底清空，只留下承重墙结构。

经过短时间的重建，这一古老的建筑以全新的方式呈现出来，室内空间风格现代。一层咖啡厅设置了一个舒适的休息区；入口处，基于棕色调的多种色彩的运用营造了一种平和的气息；座区布局也呈现多样化，既有舒适的休息区，又有专为时间紧迫的顾客设计的吧台式餐区。

二层可容纳约100人同时就餐，座区的设置与一楼相同。棕色调的运用营造出舒缓、平静的氛围，高质量的装饰材质以及墙壁上张贴的平面图案格外引人注目，吸引着顾客在皮椅上就坐。建筑立面上原有的开口被保留下来，并安装上玻璃板，这样就可以将休息区内的景象呈现出来。

1. McCafe lounge
2. Restaurant seating area
3. Order counter
4. Kitchen
5. Restroom
6. Manager office
7. Cold store
8. Storage room
9. Waste room

1. 咖啡吧
2. 就餐区
3. 点餐台
4. 厨房
5. 休息室
6. 经理室
7. 冷藏区
8. 存储区
9. 废物区

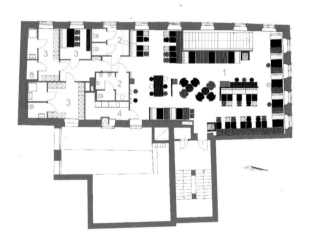

1. Restaurant seating area
2. Restroom
3. Staff
4. Waste room

1. 就餐区
2. 休息室
3. 员工区
4. 废物区

Ground floor plan　　一层平面图

First floor plan　　二层平面图

McDonald's Haus im Ennstal

麦当劳豪斯恩斯河谷地区店

Completion date: November 2009
Location: Haus im Ennstal, Austria
Designer: Viereck Architects Ltd.
Photographer: Gernot Langs, www.lanxx.at
Site area: 334,700 m²
Construction area: 50,398 m²
Seats: 82 seats restaurant + 26 seats McCafé lounge + 124 seats terrace

完成时间：2009.11
地点：奥地利 豪斯恩斯河谷地区
设计：Viereck建筑有限公司
摄影：盖诺特·朗斯（www.lanxx.at)
占地面积：334700平方米
建筑面积：50398平方米
座位：餐厅（82个）、咖啡厅（26个）、露台（124个）

This McDonald's restaurant is the first which is located directly connected to a skiing area in Styria. The rural situation required a sensitive approach to nature, thus had to be adapted to the typical McDonald's design. To meet the requirements of the Authority, a lot of wood and for the first time larch wood shingles got used at a McDonald's restaurant. This McDonald's restaurant is the first one in Austria which got realized with thermal heat pumps to optimize and reduce energy costs.

The structure in rectangular shape is a simple and straight concrete volume with a raised parapet. The façades on the upper part towards Attica run with the typical bent superior roof construction, whereby the length of the building is divided, for example, an L-bracket above the drive-in window or a scheduled pergola design with horizontal wooden slats in larch wood. This folded roof elements are neatly folded, self-contained bars, which protect against precipitation or sunlight. Particularly in the area of the terrace, which lies south-west side, these horizontal slats create a shady space and an optical loosening up to the other roof surfaces, which are on the northwest side covered with larch shingles.

The façade which faces the direction of highway, especially the terrace edge and its continuation towards the parking area will be equipped with vertical, local larch wood slats. The façade faces south (parking area) and exit will be equipped with simple laminate plates in soft surface.

1

这是第一家位于斯泰尔（Styria）滑雪度假区附近的麦当劳餐厅。独特的地理位置决定其在设计中既要保持与周围环境的和谐，又要凸显麦当劳自有的风格。为满足当局的要求，设计中大量运用了木材，这也是麦当劳集团第一次在其餐厅设计中采用落叶松木板。这也是奥地利第一家使用热力泵系统的餐厅，减少能源预算的同时，使能源利用最大化。

建筑呈现长方形造型，主体结构由混凝土打造，简约而不失特色。建筑上部立面随屋顶结构而弯曲，并在水平方向上将其分割成不同的部分，如外卖窗口上方的L支架造型以及落叶松板条材质的藤架结构。弯折的屋顶结构可以阻挡雨水的侵蚀和强光的灼晒。

室外露台就餐区位于西南侧，金属板棚架营造了一个阴凉的场所，同时与西北侧落叶松板条屋顶形成了鲜明的对比。

朝向公路一侧的立面，尤其是露台四周以及与停车场相连的一面采用竖直落叶松板条打造，而朝向南侧（停车场）的立面则采用符合材料板构造。

1.The featured logo and signage of McDonald's
2. Drive-in counter and conspicuous signpost
3. The rural location requiring a sensitive approach to nature, thus conforming the typical McDonald's design and a lot of larch wood
shingles being first used at a McDonald's restaurant
4-6. Larch wood being used as well inside and larch wooden tables and chairs creating simple yet warm environment
7. Different shades of brown creating cozy atmosphere and highlighting specialized style of McDonald's

1. 麦当劳特有的标识和标牌
2. 外卖窗口一侧以及醒目的标识
3. 独特的地理位置决定其在设计中既要保持与周围环境的和谐，又要凸显麦当劳自有的风格。设计中大量运用了木材，这也是麦当劳集团第一次在其餐厅设计中采用落叶松木板
4~6. 落叶松木材质在室内空间大量运用，木制的桌椅营造出简约却不失温馨的就餐环境
7. 不同色调的棕色营造舒适氛围，同时彰显麦当劳的专属风格

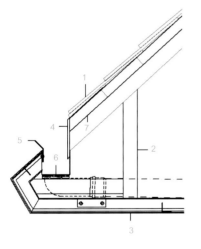

1. Shingle roof cladding
2. Steel shape galvanized 60/60/3.2
3. Supporting steel frame 130/40/2
4. Copper plate folded
5. Snowguard grid
6. Gutter
7. Substructure wood 50/50

1. 木板屋顶覆层
2. 镀锌钢板 60/60/3.2
3. 钢支架 130/40/2
4. 弯折铜板
5. 挡雪栅栏
6. （屋顶）天沟
7. 木质底部结构

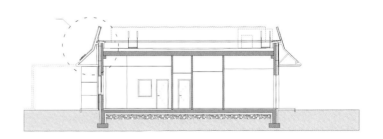

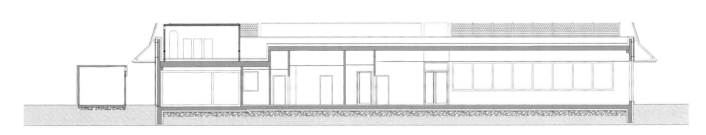

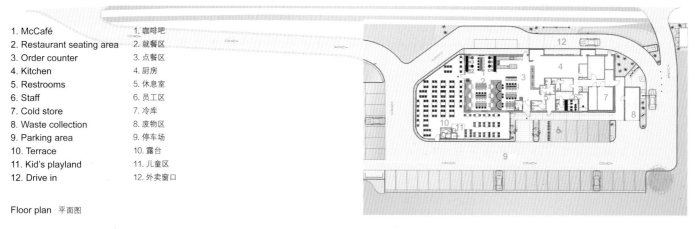

1. McCafé	1. 咖啡吧
2. Restaurant seating area	2. 就餐区
3. Order counter	3. 点餐区
4. Kitchen	4. 厨房
5. Restrooms	5. 休息室
6. Staff	6. 员工区
7. Cold store	7. 冷库
8. Waste collection	8. 废物区
9. Parking area	9. 停车场
10. Terrace	10. 露台
11. Kid's playland	11. 儿童区
12. Drive in	12. 外卖窗口

Floor plan 平面图

Burger King
汉堡王

BURGER KING stands for flame-grilled burgers, delicious fresh ingredients and crispy French-fries. Throughout the world, Burger King ensures its guests top quality service, top quality products and sparkling cleanliness. With the use of chrome, typical red leather seating and a lot of daylight thanks to the large windows, Burger King restaurants have a modern design with a classic American Diner look.

汉堡王以经营明火烘烤汉堡、炸薯条及美味新鲜食物为主，并以提供顶级服务、顶级食物和整洁的环境为宗旨。餐厅突显现代设计的同时，带有典型的美式风格，红色皮质座椅、大玻璃窗以及自然光线作为主要设计元素。

Burger King, Goldhill Centre
汉堡王金岭中心店

Completion date: July, 2011
Location: Singapore
Designer: Outofstock Design
Photographer: Outofstock Design
Area: Approx. 200 m²
Client: Burger King (Asia Pacific)

完成时间：2011.7
地点：新加坡
设计：Outofstock设计公司
摄影：Outofstock设计公司
面积：约200平方米
客户：汉堡王集团（亚太地区）

Burger King Asia-Pacific recently commissioned Outofstock to design a pilot restaurant in Singapore. The goal of the project was to create a new interior identity for Burger King. Some key points mentioned by BK was that they wanted a warm and welcoming store that would appeal to a wide audience - teens, young adults as well as families with children - the design should stand out but at the same time be accessible for the man on the street.

The designers started off by analyzing the BK brand, its advertising visuals as well as its history. They noticed that one word that kept popping up was 'flame grilled', and used this clue as a starting point. From collective experiences, their mental picture of flame grilling is closely associated with garden barbecues and camping cook-outs. These activities, often held with groups of family and friends, left indelible memories in our growing up years. This led the project to be named 'BK Garden Grill', which is based on bringing the garden, as well as colours and textures of the outdoors into the restaurant. They wanted to remind people of the joy of communal dining with family and friends in a warm and natural atmosphere, evoking memories of BBQ parties and summer camps.

The design also aimed at creating a more personable and flexible space, where potted plants can be neatly arranged or randomly placed on wooden ledges along exposed brick walls and glass windows. Framed marketing posters placed on these wooden shelves can be changed or moved about easily. More objects can be added to the fray with time as the restaurant develops its own story.

1. The specific logo of Burger King
2. The outside views being brought into the interior through the large expanse of glass
3-4. The lounge seating area set against a collage wall of materials and textures, most of which are applied throughout the restaurant, from raw concrete to clay bricks, wood veneers as well as cork, blackboard, copper and brass and BK's branding and slogans being applied in a more engaging way with this material wall
5-7. Clay, concrete and aluminum pots as pendant lamps to add to the garden atmosphere

1. 汉堡王的特色标识
2. 透过大幅的玻璃窗，可以欣赏到室外的街景
3、4. 座区后面的墙面上拼贴着各种材质，混凝土、黏土砖、薄木板、软木、黑板、红铜和黄铜等一应俱全。汉堡王品牌标识和口号被灵活地运用到这面墙壁上
5~7. 陶土、混凝土和铝材打造的灯罩营造出室外花园的感觉

Custom designed lounge seats and ottomans are upholstered with outdoor fabrics that are water repellent. The lounge seating area is set against a collage wall of materials and textures, most of which are applied throughout the restaurant, from raw concrete to clay bricks, wood veneers as well as cork, blackboard, copper and brass. BK's branding and slogans can be applied in a more engaging way with this material wall.

An overhead 'roof' trellis takes visual attention away from exposed services such as air-conditioning and kitchen exhaust trunking while also acting as cable trays for pendant lamps and spot lights. The designers suspended clay, concrete and aluminum pots as pendant lamps to add to the garden atmosphere.

They also designed simple metal framed tables whose structure is reminiscent of foldable camping furniture, but being very strong and easy to clean. They sourced for a traditional stick-back chair to complete the look of the restaurant.

汉堡王集团亚太地区总部近期委托Outofstock设计公司打造新加坡旗舰店，其目标即为营造一个汉堡王特有的、温馨热情的室内环境，以吸引不同层次顾客（青少年以及带孩子的家长）的到来。此外，他们要求这一设计在拥有别具一格的特色的同时，还会让人感到亲切。

设计师经过仔细分析汉堡王的品牌标识、广告宣传材料以及其历史之后，注意到"明火烧烤"（flame grilled）这个词出现的频率特别高，因此他们便将其作为整个设计的主线。鉴于以往的丰富经验，设计师自然而然地将"明火烧烤"与花园烧烤和露营野餐联系在一起。这些活动通常在家人或朋友之间举办，往往能够给人留下深刻而美好的印象。由此，他们将这一设计命名为"花园烧烤"，将花园、室外特有的色彩和质感引入到餐厅中来，让顾客情不自禁地联想到与家人、朋友在户外烧烤的美好场景。

设计的另一目标即为打造一个亲切、灵活的室内空间。造型多样的盆栽随意地摆放在砖墙和玻璃窗前的木质窗台上，宣传海报摆放在木质橱柜上，方便拿取。今后，这里还可以增添更多的物品，来书写餐厅自己的历史。

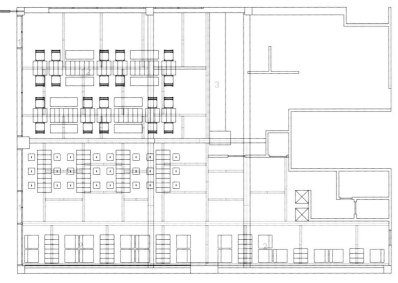

1. Entry
2. Seating areas
3. Bar counter

1. 入口
2. 就餐区
3. 柜台

定制的休闲座椅以及靠垫全部防水。座区后面的墙面上拼贴着各种材质，混凝土、黏土砖、薄木板、软木、黑板、红铜及黄铜等一应俱全。汉堡王品牌标识和口号被灵活地运用到这面墙壁上，特色十足。天花上的格架结构将空调、厨房排烟系统、电缆盘等隐藏其中。另外，设计师专门悬挂了陶土、混凝土和铝材打造的灯罩，进一步营造出室外花园的感觉。

简约的金属框桌子既可以让人情不自禁地想到可折叠露营设施，又能满足耐用、易清洗的要求。专门找到的直木条椅背座椅的运用使得餐厅的特色更加完善与鲜明。

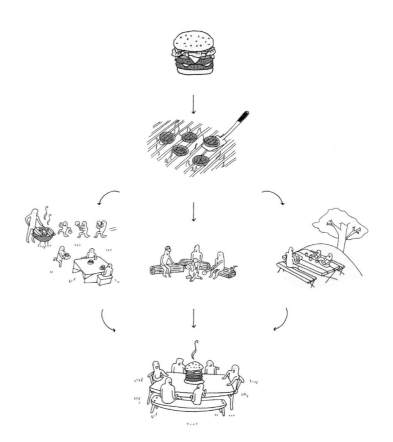

Bembos
Bembos 素食汉堡

Bembos is a Peruvian fast food chain offering hamburgers, often with Peruvian-influenced variations.

In the restaurant design, unique building form and graphics with pop tendency were important design elements. The graphics cover each room and emphasize the youth character.

Bembos 是一家秘鲁连锁快餐厅,以经营带有秘鲁特色风格汉堡食品为主。

餐厅设计中,独特的外观造型以及突显流行趋势的平面图案是主要的设计元素。平面图案遍及餐厅的各个角落,主要强调年轻与活力。

Bembos La Fontana
素食汉堡拉丰塔纳店

Completion date: August 2009
Location: La Molina, Lima, Perú
Designer: José Orrego
Photographer: Juan Solano
Area: 570m²
Project Coordination: Arq. Daniel Rondinel
Structures: Ing. Jorge Avendaño
Sanitary Installations: Ing. Cesar Torres
Mechanical and Electrical Installations: Ing. Cesar Torres

完成时间：2009.8
地点：秘鲁，利马，拉丰塔纳
设计：乔斯·奥雷戈
面积：570平方米
合作设计：丹尼尔·罗迪奈尔
结构设计：豪尔赫·阿文达尼奥
卫生装置：塞萨尔·托雷斯
机械及电器装置：塞萨尔·托雷斯

Bembos La Fontana is part of the Bembos Burger Grill franchise in Lima, Perú.

This 'big fishbowl' seems to be bigger because of its position, and gains monumentality in the urban context. Given the triangular ground and the west faced orientation, the project took advantage of the largest front, and was raised as a huge urban screen in which inner spaces and pop graphics could be shown.

This restaurant has been designed so that the service area would be compact and wouldn`t compete with public areas. As well, there's no such a thing as a 'bad place to sit'; every place has a rewarding view.

The take-out zone is suited behind the building across the parking area to optimize the parking spots.

1. Eye-catching logo and signpost of BEMBOS
2-3. The building standing like a 'fishbowl' and attracting passers-by for its unique form
4 Colours of white, blue and yellow creating a fresh and cozy atmosphere
5. Dining areas along the French window which brings outside view in
6. The order counter featuring logos of BEMBOS and playful drawings

1. BEMBOS的醒目标识和路标牌
2、3. 建筑呈现"鱼缸"造型，格外吸引眼球
4 白色、蓝色和黄色装点着整个空间，营造出清新而温和的就餐环境
5. 沿落地玻璃窗一侧的就餐区，在这里可以欣赏室外的景色
6. 点餐台彰显着BEMBOS标识和趣味十足的图案

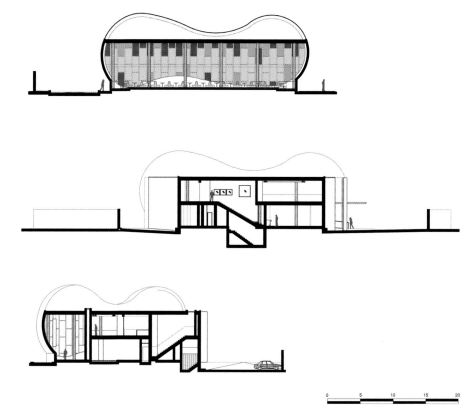

Bembos 素食汉堡拉丰塔纳店是 Bembos 汉堡集团位于秘鲁利马的分店。

餐厅朝向西侧,坐落在一块三角形地块上。鱼缸造型使其看起来更加宽敞,犹如城市背景中的一座纪念碑。

设计充分利用其巨大的立面,其看起来犹如一个城市荧幕,里面的空间和图案隐约可见。

餐厅内部空间设计巧妙,服务区紧凑而完善,与公共就餐区互不干扰。此外,就餐区每个位置都可以欣赏到美丽的景致,没有所谓的好坏座位之分。

外带区位于餐厅后面的停车场内,以便于使停车场得到充分的利用。

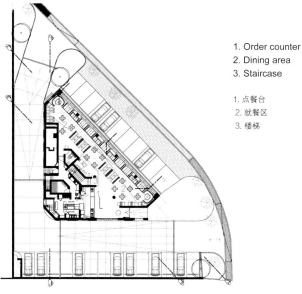

1. Order counter
2. Dining area
3. Staircase

1. 点餐台
2. 就餐区
3. 楼梯

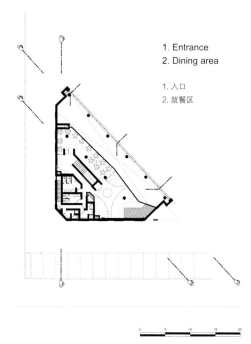

1. Entrance
2. Dining area

1. 入口
2. 就餐区

Ground floor plan　一层平面图

First floor plan　二层平面图

Bembos Larco
Bembos素食汉堡拉戈店

Completion date: 2008
Location: Lima, Peru
Designer: Arch. José Orrego

完成时间：2008
地点：秘鲁 利马
设计师：乔斯·奥雷戈

Bembos Larco is part of the Bembos Burger Grill franchise in Lima, Peru, which is notable for their iconic and carefree premise that shows a youthful attitude around the city.

The project stands out for being a building creation that contrasts with the urban landscape with a geometry like an organic device where you can see whatever is running inside.
It has two stories but the project has worked in such a way that it broke the scale and descontextualized from the surroundings and the height references. To bring that effect the ground floor has a unique big window in the lower part and the first floor has a double height space that emphasizes the entry and defines the main volume.

Inside there are sub-spaces that define different situations in the restaurant. Even when you drink a coffee, eat a burger by yourself or with your friends, play with other children or eat an ice cream, you can feel a different sensation provoked by the different spaces.

For this premise, graphics with pop tendencies that decorate all the spaces were designed to emphasize the youthful character of the project.

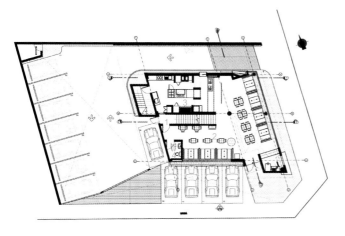
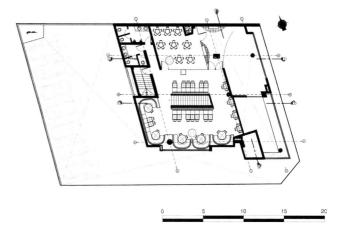

1. Entrance
2. Dining space
3. Kitchen
4. Toilet

1. 入口
2. 就餐区
3. 厨房
4. 卫生间

1-2. The iconic geometry of BEMBOS
3-6. Dining areas emphasizing casual atmosphere
7-8. Graphics with pop tendencies decorating the entire space to boast youthful character
9-10. Sub-spaces defining different situations

1、2. BEMBOS的标志性几何造型
3~6. 就餐区内营造出随意温馨的氛围
7、8. 凸显流行气息的平面图案装饰整个空间，传达出餐厅特有的活力
9、10. 不同的就餐区带有各自的特色

Bembos 素食汉堡拉戈店是 Bembos 汉堡连锁集团位于秘鲁利马的分店，因其标志性的建筑造型和独特的地理位置而著称，并诠释出一种活力十足的气息。

建筑与周围的城市景观形成鲜明的对比，几何造型的外观将室内的景象一览无余地呈现出来。
建筑共为两层，其设计打破规格和高度的制约，并从周围的环境中脱离出来。一层的底部采用宽大的玻璃窗打造，二层采用双高空间设计，使得入口和主体结构更加突出。
店内被分割成若干个小空间。在这里，或是喝杯咖啡，吃个汉堡，或是与朋友小聚，与孩子玩耍，都能感受到不同空间营造的独特氛围。

展现流行趋势的平面图案用于装饰整体空间，进而强调出餐厅独有的年轻与活力。

Nando's
Nando's 连锁快餐厅

Nando's is a chain of restaurants that began in South Africa in 1987 whose specialty is Portuguese-style chicken. Their casual style and ability to adapt to different markets with their creative branding made them the success they are today. The company is a big supporter of the South African art scene and one can see work displayed on the walls and on the wire mesh grates that hang all around.

Maintaining individuality of the restaurants is part of Nando's brand identity. Each Nando's restaurant has its own character based on the practicalities of the site and on the people who will be visiting. Strong, natural materials are commonly used, from marble to slate to antique oak, and as much natural light as possible.

Nando's 连锁快餐厅在 1987 年始于南非，特色食物即为"葡国鸡。"随意的风格以及适应不同市场需求的创意推广方式为今天的成功奠定了基础。Nando's 是南非艺术的支持者，在其餐厅墙壁上或网格结构内可以看到展出的各种画作。

Nando's 一直注重保持独特性。每一家餐厅由于选址和顾客类型不同，都被赋予各自的特色。天然材质如大理石、石板、橡木等被广泛选用。自然光线也被视作一种自然设计元素。

Nando's Dublin

Nando's都柏林店

Completion date: December 2011
Location: Mary Street, Dublin, Ireland
Designer: Buckley Grey Yeoman
Photographer: Hufton + Crow
Client: Nando's Chickenland plc
Awards : Finalists, 2012 Interior Architect of the Year

完成时间：2011.12
地点：爱尔兰，都柏林，玛丽街
设计：伯克利·格雷·约曼设计公司
摄影：Hufton + Crow 摄影
客户：Nando's快餐连锁集团
获奖：2012年度室内建筑奖决赛入围

Buckley Grey Yeoman designs the striking new Dublin restaurant for Nando's.

London-based architecture practice Buckley Grey Yeoman (BGY) has completed its latest restaurant for the Nando's chain on Dublin's Mary Street in the city centre. Having completed projects for Nando's branches in London, Manchester and Glasgow, BGY was given a flexible brief that revolved around bringing a fresh design approach to Dublin and delivering a restaurant unlike any other Nando's.

The new restaurant welcomes customers with a design that encapsulates the warm, fun atmosphere of the Nando's brand whilst also showcasing BGY's approach to creating carefully detailed spaces. As customers enter the restaurant, they are greeted by a servery area, clad at random intervals with exposed reinforcement bars, a material traditionally concealed within concrete, to create a ribbed effect. A blend of up and down-lighting is used to bring this feature to life and transform its appearance as the light conditions change throughout the day.

Whilst walking through the 80-seat restaurant, customers experience a range of vibrant colours that are accentuated by industrial-looking fixtures and fittings. From brightly coloured hand-made wall tiles to reclaimed industrial furniture, an eclectic range of materials and a lively palette of colours have been selected by BGY to animate the space.

A prominent example of the high-level of detailing found throughout the project is the intricate ceiling, made out of hand-woven English willow that appears to float above diners. At the back of the restaurant, the tactile willow panels form a backdrop to the booth-seating area and are low enough to be touched while from the outside the ceiling creates an inviting gesture to the street.

Paul Thrush from Buckley Grey Yeoman, said: 'The range of materials, textures and colours in the restaurant speaks volumes about the project's adventurously open brief. For each restaurant we've delivered for the Nando's, the client has encouraged us to do something new and take a fresh approach, with all projects linked by a sense of fun and keen eye for detail.'

1-4. The intricate ceiling, made out of hand-woven English willow appearing to float above diners
5-8. Special lighting bulbs suspended from the ceiling to create a warm atmosphere
9. Brightly coloured hand-made wall tiles used to animate the space
10. The servery area, clad at random intervals with exposed reinforcement bars, a material traditionally concealed within concrete, to create a ribbed effect

1~4.精美的天花,运用当地编织技术和柳条手工制作而成,犹如悬浮在就餐顾客的头顶上方
5~8.独特的灯饰从天花上悬垂下来营造出温馨的氛围
9.色彩亮丽的瓷砖用于装饰墙面,增添了空间活力
10.由间距不等的竹节钢筋(一种由水泥打造的材质)饰面的点餐台,打造出罗纹效果

伯克利·格雷·约曼设计公司为Nando's集团打造了一个引人瞩目的都柏林分店。

伦敦设计公司伯克利·格雷·约曼完成了Nando's位于都柏林中心区玛丽街分店的设计,之前曾设计了Nando's位于伦敦、曼切斯特和格拉斯哥的分店。此次,客户要求打造一个全新的设计,与Nando's其他分店要大有不同。

温馨而趣味性十足的空间欢迎着顾客的到来,体现细节的设计元素同样被融合进来。顾客走进餐厅,首先映入眼帘的是由间距不等的竹节钢筋(一种由水泥打造的材质)饰面的点餐台,打造出罗纹效果,在灯光的闪耀下更加生动,还可以随着光线的变化不断改变外观。就餐区共包括80个座位,鲜亮的颜色在工业风格的家具和设备的衬托下格外吸引眼球。墙壁上亮色的手工瓷砖到回收的工业风格家具,所有的元素都是设计师精心选择的。

餐厅空间中最能体现细节设计的当属精美的天花,运用当地编织技术和柳条手工制作而成,犹如悬浮在就餐顾客的头顶上方。餐厅后侧,大幅柳条板为就餐区打造了独特的背景。从外面看进来,巧妙设计的天花格外引人注目。

来自于伯克利·格雷·约曼设计公司的设计师保罗·特拉什这样评价这个项目:"我们选用的材质、颜色充分诠释了大胆灵活的设计理念。每次为Nando's设计,他们都会要求增添新的元素或者采用新的方式。Nando's所有的项目是通过趣味性和细节联系在一起的。"

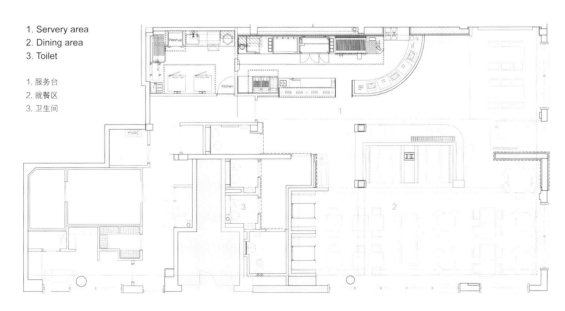

1. Servery area
2. Dining area
3. Toilet

1. 服务台
2. 就餐区
3. 卫生间

Floor plan 平面图

Nando's Ashford

Nando's阿什福德店

Completion date: December 2011
Location: Ashford, Kent, UK
Designer: Blacksheep
Photographer: Ben Webb
Area: 380 m²
Client: Nando's

完成时间：2011.4
地点：英国，肯特，阿什福德
设计：Blacksheep室内设计公司
摄影：本·韦伯
面积：380平方米
客户：Nando's连锁快餐集团

South African heritage meets New York loft style.

Having previously carried out two refurbishments for the restaurant chain Nando's (in Soho and Southampton), Blacksheep was commissioned to create a site-specific design for its new branch in Ashford, Kent.

The brief was to design a concept space that steered the client away from the traditional Nando's design and put a Blacksheep signature style and twist on the well known and loved chicken restaurant. The client wanted to attract a more mid market customer base and move away from a mass market centred business structure.

The client had a clear idea of audience, however the challenge was that the location was in a remote leisure park that already attracts a certain type of cliental and has a variety of competing restaurants in the vicinity. Secondly, the building, formerly one half of a nightclub, was a vast, cavernous shell, which had too much space. The height was reduced from 10m to 7m to form a more workable area, this was accomplished by a suspended wooden lattice that created a boundary around the dining area and low ceiling lights to create a sense of intimacy.

Blacksheep's first priority was to attract passing trade. This was achieved by opening up the frontage and creating an eye-catching façade using diagonal strips of oak accompanied by metal outdoor furniture with Iroko table-tops.

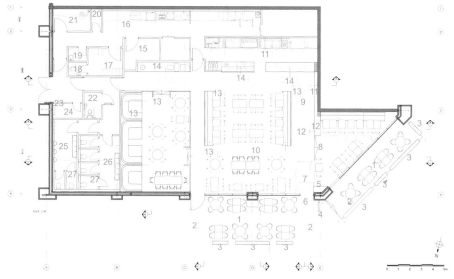

1. External seating area		1. 室外就餐区
2. Awning		2. 标牌
3. Planter		3. 植物
4. Projetcting sign		4. 凸出标识
5. Fire Panel		5. 防火板
6. Main entrance		6. 主入口
7. Entrance lobby		7. 入口大厅
8. Step up		8. 台阶
9. Waiting area		9. 等候区
10. Restaurant		10. 餐厅区
11. Condiments shelf		11. 调料橱柜
12. Meet & greet		12. 接待区
13. Banquette		13. 长椅就餐区
14. Condiments & drinks		14. 调料及饮品区
15. Dry store		15. 干货存储区
16. Wash up		16. 奥洗区
17. Staff room		17. 员工区
18. Staff toilet		18. 员工卫生间
19. Cleaners cupboard		19. 清洁用品橱柜
20. General store		20. 普通存储区
21. Office		21. 办公区
22. Disabled toilet		22. 残疾人专用卫生间
23. Distribution board		23. 配电箱
24. Electrical/Services cupboard		24. 电器柜
25. Male toilet		25. 男士卫生间
26. Female toilet		26. 女士卫生间
27. Fire exit		27. 消防出口

1. Bespoke light bulb suspended from the ceiling
2. Wooden wall and table creating a warm atmosphere
3-4. Dining area
5. Large-scale bespoke artwork, in the style of Southern African Ndebele tribe, made of thousands of painted wooden dowels
6-7. 3D artwork wall
8-9. Distinctive lighting fixtures

1. 定制的灯饰从天花上悬垂下来
2. 木板墙壁和桌子共同营造温馨的氛围
3、4. 就餐区
5. 数千块的木板打造了一个南非尼贝利（Ndebele）部落特有的艺术品，挂在那里欢迎着顾客的到来
6、7. 三维艺术装饰墙
8、9. 独特的灯饰

Taking into account the scale of the interior – 10m high ceilings and a total area of 380 sqm – Blacksheep decided to give the restaurant a New York loft feel, combining wood and metalwork accents with a pared-down aesthetic. To make the space feel more intimate, Blacksheep installed a raised platform with booth seating and bespoke timber latticework, as well as designing huge custom-made light-shades. Blacksheep also created a large-scale bespoke artwork, in the style of Southern African Ndebele tribe, using thousands of painted wooden dowels. Bold and iconic, it hangs in pride of place as you enter the restaurant. The 192-cover restaurant opened its doors to the public in December 2011.

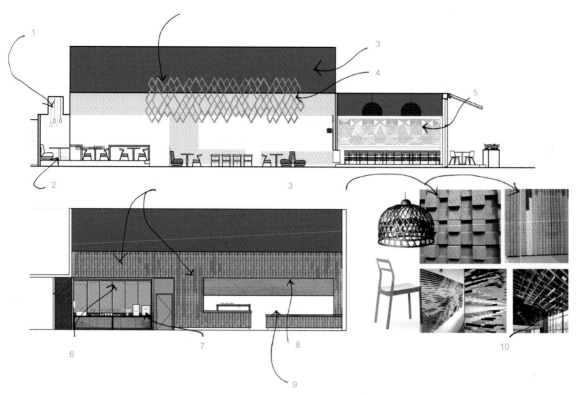

1. Groups of pendants hung from within bulkhead
2. Booths
3. Upper walls & ceiling painted dark charcoal grey
4. Structure in light oak
5. 3D artwork wall
6. Waxed mud steel paneling with Nando's graphics applied
7. Oak & painted timber table
8. Oak clad facia
9. Granite work tops with hardwood lipping at front
10. Main ceiling frame – oak frame below dark charcoal ceiling colour

1.嵌入舱壁的小吊灯
2.就餐区
3.炭灰色天花
4.橡木结构
5.三维艺术墙
6.带有Nando's平面图案的打蜡钢板
7.橡木桌子
8.橡木招牌
9.花岗岩饰面及实木镶边柜台
10.天花特色结构——橡木框架从炭灰色天花上悬垂下来

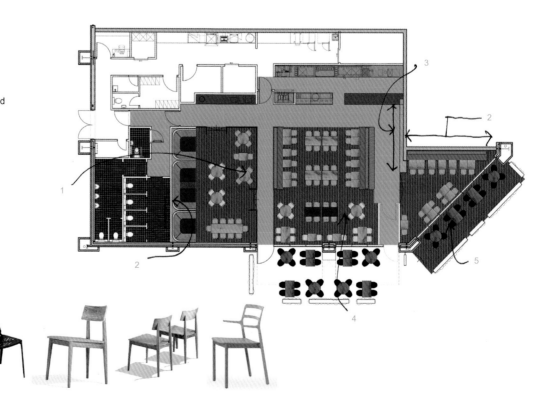

1. Loose seating at platform edge
2. 3D artwork wall
3. Paintings on mesh frame
4. Revised seating area layout with additional covers
5. Black coloured external chairs and timber table tops

1.平台边缘的就餐区
2.三维艺术墙
3.网架上的装饰画
4.就餐区
5.黑色座椅和木面桌子

南非传统文化与纽约阁楼风格的融合

Blacksheep 室内设计公司之前曾负责 Nando's 其中两家餐厅的改造项目,这一次受托为其位于阿什福德的新店打造一个具有地区特色的设计。

设计主旨即为打造一个全新的概念空间,摒弃 Nando's 的传统风格,赋予其 Blacksheep 特色,同时更要与众所周知的炸鸡店大不相同。客户的目标顾客群指向中端市场,而不是以大众市场为主体。

首先,在目标顾客群定位明确的基础上,设计师仍然面临的挑战是:餐厅选址在一个位置偏僻的休闲公园内,其顾客群已相对固定,而且周围有很多其他餐厅。其次,建筑的一部分之前曾被用作夜总会,外表普通而内部空旷。为解决这两个问题,设计师首先打造了一个极为引入注目的立面,对角线排列的橡木图案与衣罗可木饰面的室外金属材质家具相互呼应。

室内总面积 380 平方米,屋顶高达 10 米。设计师在充分考虑了室内规格之后,决定打造纽约阁楼风格,将木材和金属元素结合在一起。为营造更多的亲切感,设计师在这里引进了一个上升平台,上面设置着座区和木格子结构。灯具也是专门定制的。此外,他们还利用数千块的木板打造了一个南非尼贝利(Ndebele)部落特有的艺术品,挂在那里欢迎着顾客的到来。餐厅可容纳 192 人就餐,并于 2011 年 12 月开始营业。

1. 3D artwork wall
2. Leather upholstered booths
3. Wasters of small pendants fixed within bulkhead
4. Timber clad bulkhead
5. Metal clad doorway
6. Paintings hung on artwork mesh display racking
7. Timber flooring turns up wall
8. Oak shelf and crafter station
9. Custom feature pendants
10. Timber framed glazed vestibule
11. 3D artwork proposal: coloured pegs define large-scale image
12. Artwork display rack
13. Booth pendants
14. Custom pendants

1. 三维艺术墙
2. 皮面座椅
3. 嵌入舱壁的小吊灯
4. 木材饰面舱壁
5. 金属饰面入口
6. 网架上悬挂的装饰画
7. 木质地面一直延伸到墙壁上
8. 橡木橱柜
9. 定制特色吊灯
10. 木框玻璃门廊
11. 三维艺术墙设计方案：彩色钉状物排列成一幅巨大的画面
12. 艺术品展示架
13. 座区上方悬垂的灯饰
14. 定制灯饰

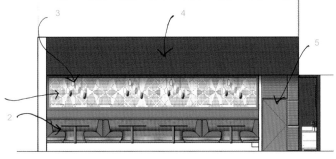

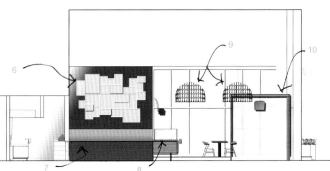

1. Timber effect tiled walls
2. Deep blue coloured mirror frame
3. Vanity trough clad in glazed white ceramic tiles & white Corian
4. Painted walls
5. Timber rail
6. Powder coated pendant hauts
7. Coloured mixer tap

1. 犹如木质一般的瓷砖墙壁
2. 深蓝色镜框
3. 白色抛光瓷砖和白色可耐丽饰面水槽
4. 喷漆墙面
5. 木扶手
6. 喷粉灯罩
7. 彩色水龙头

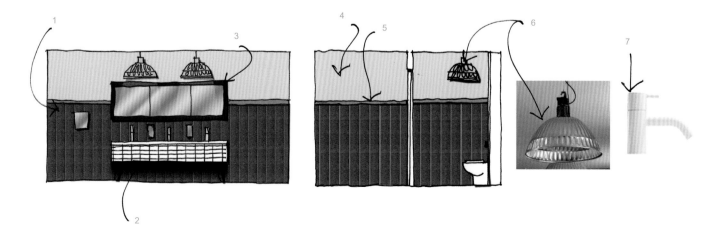

9

Nando's Canberra
Nando's堪培拉店

Completion date: 2011
Location: Canberra, Australia
Designer: Design Clarity
Principle Designers: Kerstin Braun, Rebecca Boland
Photographer: Design Clarity
Area: 135m²

完成时间：2011
地点：澳大利亚，堪培拉
设计：Design Clarity室内设计公司
主设计师：克尔斯汀·布劳恩、瑞贝卡·伯兰
摄影：Design Clarity室内设计公司
面积：135平方米

Explorer concept sets hearts on fire in our nation's capital. The brief from the client was to design a unique space within the Nando's family which is true to the their natural, raw and solid design language and intent. The concept for this restaurant was inspired by the history of Portuguese sailors exploring the world, which led to a very solid and nostalgic design with nautical influences.

As always with Nando's the challenge is to comply with the design brief and come up with a completely new look with different design elements not seen in any other previous Nando's site.

The finishes include tone-on-tone decorative floor and wall tiles, white washed brick, white ceiling with raw timber joinery, steel pins and copper details, to create natural aged, raw look.
The long gun-barrel site footprint inspired the ship's hull structure. The feature of the restaurant is the repetition of the architectural curved plywood panels located above the raised seating area with rope and copper details. These panels evoke and abstractly convey the interior structure or 'ribs' of an explorer's sailing ship. Raw and laser cut timber branding features as well as rustic wall paint graphics lead the visitors into the Nando's world and their many stories about Peri Peri and the Legend of the Barcelos Cockerel.

1. Raw and laser cut timber branding feature
2-3. A very solid and nostalgic space with nautical influences and feature of the restaurant of the architectural curved plywood panels located above the raised seating area with rope and copper details
4-5. Order counter
6. The brown colour bringing warm feeling
7. Wall of white tiles creating a natural and aged look
8. Rustic wall paint graphics leading the visitors into the Nando's world

"探索"点燃热情

客户要求打造一个独特的空间，并融合Nando's自然、纯朴、真实的设计语言。这一设计灵感源于葡萄牙航海者探索世界的历史，从而营造了一个真实而怀旧的空间，并带着些许的海上情怀。

这一设计中面临的主要挑战即是在遵循Nando's理念的同时，运用其之前未曾用过的各种设计元素，营造出一个全新的形象。

装饰元素包括同色地面、白色瓷砖墙、漂白砖墙、木工、钢钉及铜材质装饰的白色天花，旨在营造一种自然仿旧的效果。

长枪管式的空间形状促使设计师决定将餐厅打造成船壳造型。这里的主要特色即为抬升的座区上方由绳索和铜板固定的弯曲胶合板结构，不断重复构建出独特的造型。这些结构让人不禁联想起航海船的内部结构。激光切割的原木标识板与墙面上的平面图案引领着顾客走进Nando's的世界，在这里可以听到关于辣椒酱和巴塞罗斯小公鸡的传奇故事。

1. 激光切割的原木标识板
2、3. 一个真实而怀旧的空间，并带着些许的海上情怀。座区上方由绳索和铜板固定的弯曲胶合板结构，不断重复构建出独特的造型
4、5. 点餐台
6. 棕色调营造出温馨的氛围
7. 白色瓷砖墙壁诠释出自然而怀旧的气息
8. 墙面上的平面图案引领者顾客走进Nando's的世界

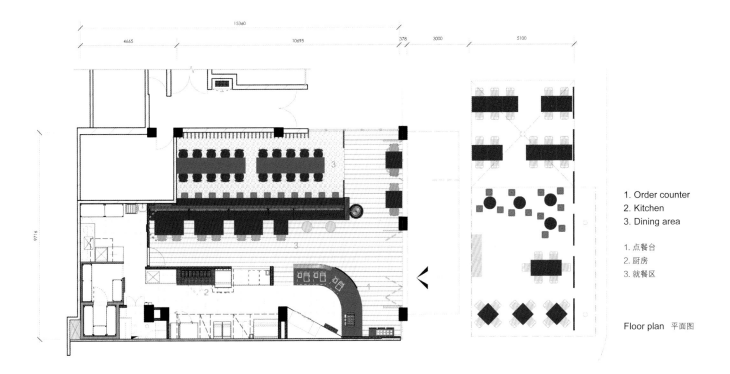

1. Order counter
2. Kitchen
3. Dining area

1. 点餐台
2. 厨房
3. 就餐区

Floor plan 平面图

Pizza Hut
必胜客

Pizza Hut is an American restaurant chain and international franchisethat offering different styles of pizza along with side dishes including salad, pasta, buffalo wings, breadsticks, and garlic bread.

Corporately known as Pizza Hut, Inc., it is a subsidiary of Yum! Brands, Inc., the world's largest restaurant company. As of 2012, there were more than 6,000 Pizza Hut restaurants in the United States, and more than 5,139 store locations in 94 other countries and territories around the world. The goal in the restaurant module is to create a comfortable ambiance for friends and family to share pizza in a relaxed, fun atmosphere.

Pizza Hut（必胜客）是美国一家连锁快餐集团，提供不同风味的比萨以及沙拉、意大利面食、布法罗炸鸡翅、面包条、大蒜面包等。

如今，Pizza Hut, Inc 隶属于 Yum! Brands（百盛餐饮集团，世界最大的餐饮公司）。截至 2012 年，美国共有 6000 多家必胜客餐厅，在全球 94 个国家内拥有 5139 家分店。必胜客餐厅设计的模式即为营造一个舒适的就餐氛围，供朋友和家人享用美味比萨。

Pizza Hut Prime Time

黄金时段—必胜客

Completion date: June 2011
Location: Touchwood, Solihull, UK
Designer: Paul Winch-Furness
Photographer: Paul Winch-Furness
Client: Pizza Hut UK
Area: 230m²

完成时间：2011.6
地点：英国，索利哈尔，塔奇伍德
设计：保罗·温切尔-弗内斯
摄影：保罗·温切尔-弗内斯
客户：必胜客英国集团
面积：230平方米

The brief from Pizza Hut UK was to develop a restaurant around the concept of a more sociable, dramatic and atmospheric restaurant that encourages experimentation, customization, namely 'trying new things'.

Based on the idea of 'Prime Time' when family and friends gather round on a Saturday night watching TV together, enjoying the shared experience, the designers developed the concept as a way of enveloping the kind of social energy that should run through the interior, service, menu and various brand touch-points. The 'Prime Time' concept has since influenced all Pizza Hut UK's marketing and promotional projects.

Before the design was even started, in depth consumer research was undertaken, which identified the complex and subconscious process that consumers go through in choosing a restaurant and all the elements that they experience throughout the customer journey in the restaurant.

This prompted Pizza Hut UK to refresh its whole customer experience which started with the overhaul of the menu to offer what Sanjiv Razdan, Pizza Hut UK's Brand Development Director, calls 'craveable' food. The thread that runs throughout the new-look menu is customization, and the freedom for customers to tailor any dish to personal taste.

Another change is replacing Pizza Hut UK's longstanding pizza and pasta buffet with 'Pizza Parade' as a way of bringing more theatre to the restaurant and more personal interaction between customers and staff. 'Paraders' or 'choreographed servers' as Razdan calls them, will visit tables with freshly cooked pizzas and pasta which customers can take from as often as they like providing a fun new way to eat. The hope is that it will encourage customers to try new things and, with lively, chatty service encouraged, will make for an interactive experience, maximizing sociability.

Customers' first glimpse of the new 'Prime Time' concept has now been unveiled at Touchwood, Solihull. The design of the restaurant façade and entrance has been evolved with an exterior that features a multicoloured banding replicated from the redesigned menu. The brand red beacon at the entrance features internal movement and light. Adjacent to the dramatically lit red entrance tunnel and Pizza Hut UK's 'Eat Drink and Enjoy' message projected onto the mall, it helps add a sense of theatre and beckon customers into the restaurant.

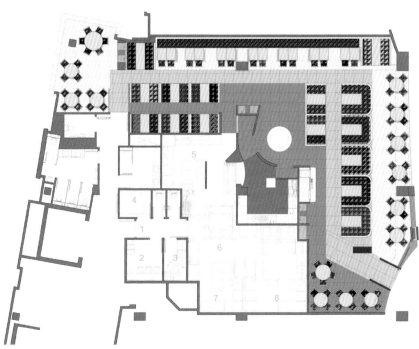

1. Lobby
2. Team room
3. Office
4. Plant
5. Dishwash area
6. Kitchen area
7. Freezer
8. Chiller

1. 大厅
2. 员工休息区
3. 办公区
4. 机械区
5. 餐具清洗区
6. 厨房
7. 冷冻区
8. 冷藏区

Floor plan　平面图

1. The colourful framed graphics on the wall adding interest to the booth
2. The red lighting fixtures dangling above the salad bar
3. Textured acoustic panels lining the back wall, with a patterned gobo light effect running gently over them, creating a subtly theatrical ambiance
4. Ice cream factory
5. Abstract art in the form of 'artful' pizza pans which have been backlit
6. The Salad Bar

1. 就餐区墙壁上悬挂着的不同颜色的镶框平面图案增添了空间趣味性
2. 红色的灯饰悬挂在沙拉餐台上方
3. 吸音板沿着后面的墙壁排列，灯光投射出来的花样图案映在上面，营造出微妙的舞台氛围
4. 冰激凌工坊
5. 比萨炉形状的艺术装饰结构，采用背光照明
6. 沙拉吧

Once in the restaurant, the salad bar forms the centre stage, promoting Pizza Hut UK's free unlimited salad with every main course. The area is bright and fresh, simply presented, with a halo of dramatic 'stage' lights overhead ensuring the food is the main star here. Curved walls provide a backdrop for the salad bar, and Pizza Parade staff appear (theatrically) from the stage 'wings', making their entrance into the restaurant.

Booth and banquette seating surrounds the salad bar. The booths are designed to allow customers to feel the 'buzz' of the restaurant, but feel contained in their own private space.

Seating in contrasting colours defines areas of the restaurant; half of the booths use seating in rich bronze with lime green piping providing a zing of accent colour from the new palette; adding warmth to the restaurant, the remaining booths are covered in dark plum with some featuring large red light shades and are partially screened from neighbouring booths, making for a more intimate space.

On the far side of the restaurant banquettes provide a more flexible seating area accommodating tables for two people which can be grouped together for larger parties. This more open dining area is separated out from the rest of the restaurant through the use of a wooden floor and ceiling giving the area its

own identity. Textured acoustic panels line the back wall, with a patterned gobo light effect running gently over them, creating a subtly theatrical ambiance.

Towards the front of the restaurant is the ice cream factory, where children can have fun creating their own dessert. With three children's sprinklers, two adult spices and a variety of sauces, parents and children can interact together, sharing the family experience. They also designed a new menu, where the design and colour palette plays an especially strong influence on the graphics in restaurant; from the restaurant's exterior treatment to the colourful frames featuring 'artful' pizza pans with abstract laser cut designs lining one wall of the restaurant and the playful graphics which are a light hearted take on well known phrases in colours from the menu.

The new concept puts much greater emphasis on the customer, through both the service and the environment ensuring that they have a great and memorable experience. Razdan wants Pizza Hut UK's restaurant customers to feel 'uninhibited, embraced and light-hearted.' Releasing inhibitions means making customers feel confident that they are in a safe, nurturing place where they know not only that they will have a good time but that they can try new things without risk. Feeling embraced should mean that every element of a family enjoys a visit to Pizza Hut UK, and Razdan cites the examples of films like Finding Nemo or Shrek as examples of the universal, warm appeal that the chain is seeking. Light-hearted means helping people to have fun and enjoy everyday moments: 'We know we're not about a once-in-a-lifetime experience — we're about little treats.'

This concept is currently being implemented into ten test bed sites throughout London and the Midlands, with a view to rolling out the concept nationally.

客户要求打造一个适于社交、氛围融洽的独特餐厅，鼓励顾客体验尝试新鲜事物。

设计理念源于"黄金时段"这一概念。全家人在周六的夜晚围坐在一起看电视、分享经历的场景是这一理念形成的基础，设计师因此构思了一种方式，即将这种社会能量渗透到餐厅中的服务、菜单、品牌标识等各个方面。"黄金时段"的理念对英国的所有必胜客餐厅的营销和推广都起到了一定的作用。

设计开始之前，设计师对于与顾客相关的内容做了深入的调查。从他们如何选择一个餐厅到餐厅中经历的一切。调查促使必胜客对于顾客的整个就餐经历进行了调整，从菜单的改变开始。必胜客品牌发展总监提出一种"渴望食品"，即顾客可以根据自己的口味要求定制食品。

另外一个改变即为采用"比萨游行"的模式取代原来的固定自助餐台，不仅

5

增添了趣味性,更能促进顾客和员工的交流。餐厅员工将新做的比萨或各种面食端到顾客面前,让他们自助选择。这样可以鼓励顾客品尝新品种,带来互动式体验。

"黄金时段"理念的必胜客餐厅首先出现在索利哈尔郡(Solihull)的塔奇伍德(Touchwood)。餐厅外观和入口经过仔细翻修,以改进的菜单为原型的多色标识牌构成了餐厅的外部特色。入口处的红色灯标增添了内部的流动性,同时与柜台上印着的"Eat Drink and Enjoy"("吃喝玩乐")的标识语一起营造出一种剧院的氛围,指引着顾客走进来。

走进餐厅,首先映入眼帘的便是萨拉自助餐台,无限量提供与主餐搭配的免费沙拉。这一区域明亮而简洁,头顶上的舞台灯照明确保使食物成为主角。蜿蜒的墙壁营造了背景,餐厅负责"比萨游行"的工作人员从两侧等待入场。

就餐区围绕在沙拉自助餐台的四周。餐区的设计让顾客既能感受到餐厅内的热闹范围,又能在自己的小空间中独享乐趣。不同颜色的座椅将餐厅划分开来,一部分座椅采用青色装饰,带来温暖的气息,另一部分采用深紫色装饰,给人亲切之感。

餐厅另一侧的长椅就餐区则更为灵活,供两人就餐的桌子可以组合在一起,供团体聚会使用。这一区域更加开放,地面和天花采用木材装饰,具有自己的特色并与其他空间分隔开来。吸音板沿着后面的墙壁排列,灯光投射出来的花样图案映在上面,营造出微妙的舞台氛围。

"冰淇淋工坊"位于餐厅的前方,孩子们可以自己动手制作甜点。父母可以与孩子们互动,享受美好的家庭氛围。此外,他们还定制了新的菜单,其设计和色彩对于餐厅内的平面设计带来很大的影响,例如餐厅外立面处理、墙壁上带有比萨盘图案的相框以及趣味十足的各种平面图案。

新的设计理念着重突出顾客,无论从服务还是环境上确保为他们带来一个美好而难忘的就餐经历,让他们感到随意、关爱、放松。随意即是富有安全感,不仅能够享受时光,还能够品尝到新的食物;关爱是指家庭中的每一位成员都愿意来到这里,犹如电影《海底总动员》和《怪物史莱克》中所体现的主题。放松意味着人们可以在这里享受每一刻,这虽不是一生仅有一次的经历,但却让人难忘。

这里理念已被运用到伦敦及英国中部地区的10家必胜客餐厅中,并将进一步被引进到全国各地。

PizzaExpress
PizzaExpress 快餐厅

PizzaExpress is a restaurant group with over 400 restaurants across the United Kingdom and 40 overseas in Europe, Hong Kong, India and the Middle East. It was founded in 1965 by Peter Boizot. In 2002, PizzaExpress launched PizzaExpress Prospects Contemporary Art Prize with pop artist Peter Blake. Peter Blake's connection with PizzaExpress was extended when he donated 26 original pieces to the Chiswick restaurant. PizzaExpress created a 'Living Lab' in October 2010, in Richmond, trialing new ideas from design to sound, collaborating with designer Ab Rogers.

In 2011, PizzaExpress launched a major rebrand of its UK restaurants, with menu changes, a black and white logo and the widespread use of stripes, both for staff uniforms and for restaurant decor.

PizzaExpress 餐饮集团在英国共有 400 多家店面，并在欧洲、香港、印度及中东等 40 多个地区设有分店。2002 年，PizzaExpress 与流行艺术家皮特·布莱克（Peter Blake）合作举办"PizzaExpress 当代艺术奖"（PizzaExpress Prospects Contemporary Art Prize）。此后，皮特向其位于奇西克（Chiswick）的分店捐赠了 26 件原创作品。2010 年 10 月，PizzaExpress 与设计事务所 Ab Rogers 合作在里士满打造了名为"体验实验室"的分店，从空间设计到音响设备寻求了一种全新的方式。

2011 年，PizzaExpress 位于英国的店面重新设计了标识和菜单，黑白相间的标识以及条状结构被运用到员工制服和空间装饰上。

Living Lab
体验实验室—PizzaExpress快餐厅里士满店

Completion date: 2010
Location: London, UK
Designer: Ab Rogers Design, in collaboration with DA Studio
Photographer: Ab Rogers
Client: PizzaExpress

完成时间：2010
地点：英国，伦敦
设计：AB-罗杰斯设计（DA工作室合作设计）
摄影：AB-罗杰斯
客户：PizzaExpress快餐厅

Living Lab: Ab Rogers Design creates innovative concept restaurant for PizzaExpress.

In summer 2010, PizzaExpress commissioned London design agency Ab Rogers Design (ARD) to transform its Richmond restaurant into a 'Living Lab' combining experiential design, radical menu changes, and advanced developments in restaurant acoustics. With an innate love of Italy and its inimitable food – coupled with fond childhood memories of visiting his local PizzaExpress with his Italian grandmother – Rogers jumped at the chance.

Placing an open kitchen at the heart of the restaurant, the joy of food and the theatre of pizza-making become the focus of ARD's design. The vibrant drama of pizza being tossed and pummelled into life is encircled by a red ribbon of activity, with seating booths, and bar encompassing the open kitchen.

Inspired by traditional Neapolitan open-air stalls, a new kiosk punches through the restaurant wall, allowing passersby to get food on the go. The culture of Naples permeates throughout the new restaurant, and can even be heard in the toilets, where atmospheric recordings of Neapolitan life are played.

ARD's collaboration with Graphic Thought Facility, brings a fresh energy to the restaurant's graphic scheme with dynamic new colour, signage, uniforms and table settings.

Rogers and his team alongside leading acoustician Sergio Luzzi composed a perfectly balanced soundscape. Acoustic circles hanging dramatically from the ceiling, and bespoke domes, suspended over intimate booths, create a series of private spaces within a buzzing public place. The domes are fitted with iPod docks, dimmer switches and call buttons, to customize the environment. The acoustics have been widely commended, not least by BBC TWO's See Hear programme.

The Living Lab redesign includes activity areas for children, museum-esque interactive 'stealth-learning' video games and a communal drawing table where kids can also make their own pizzas, assisted by the pizzaiolos. Elements of ARD's ground-breaking scheme will expand to PizzaExpress branches nationwide.

体验实验室：AB-罗杰斯设计公司为PizzaExpress快餐连锁店打造了一个全新概念的餐厅。

2010年夏天，PizzaExpress快餐厅委托AB-罗杰斯设计公司将其位于里士满的餐厅改造成一个"体验实验室"，具体工作包括空间改造、菜单更换以及音响系统更新。罗杰斯非常热爱意大利这个国家以及那里的美食，加上对童年与祖母一起在PizzaExpress就餐的美好回忆，他欣然接受了这项工作。

他首先在餐厅中央打造了一个开放式厨房，由此一来，pizza的制作过程便犹如一场戏剧表演成为整个设计的中心元素。蜿蜒的红色吧台将开放式厨房围和起来，这里上演着一场场精彩的演出。

深受那不勒斯传统风格的露天餐厅影响，设计师将餐厅原有的一面墙壁凿开并打造了一个小的售货亭，这样路人便可以随时买到食物。那不勒斯文化在餐厅中随处可以感受到，即使在卫生间内都可以听到关于那不勒斯生活的音乐。

设计师与伦敦著名的平面设计机构（Graphic Thought Facility）合作，运用新的色彩、标识以及餐具，为餐厅注入了更多的活力。

设计团队还和著名的音响师塞吉奥·鲁兹（Sergio Luzzi）共同打造了一个完美而平衡的音响范围。从天花上悬垂下来的吸音圆盘以及悬浮在休息区上方的圆顶结构将整个空间分割开来，形成一系列的单独私密空间。圆顶结构内安装有音响播放器、调节开关以及呼叫按钮，巧妙的设计定义了空间整体氛围。这里的音响系统曾受到广泛的好评。

此外，设计还包括儿童活动区的打造，互动式的视频游戏以及供孩子们画画使用的长桌别具特色。当然，孩子们还可以在辅导之下，自己动手制作比萨。这一全新的设计理念将被运用到PizzaExpress世界各地的餐厅中。

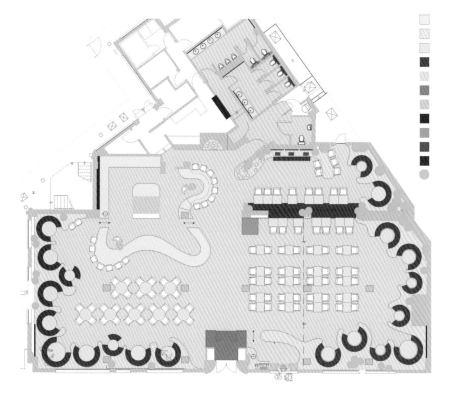

Corian Workop (See Material Schedules For Options)
Booth Top Surface : Abet Laminati Solidcore, White
Free Standing Tables
Fixed Booth Seating
Floor Tiles
Entrance Matt
Glass Food Display To Show To The Street Side
Wall Mounted Shelves For Wines
Glass Fridge
Screens For Interactive Games
Food Display, Glazing From The Toilet Corridor
Herbs

可丽耐操作台
就餐区表面装饰材质（Abet Laminati 公司提供）
可移动餐桌
固定就餐区
地砖
入口亚光材质地面
玻璃食品陈列柜，便于向街上的行人展示餐厅内商品
盛放酒品的壁柜
冰箱
互动游戏屏幕
食品展示
植物

1. The vibrant drama of pizza being tossed and pummelled into life encircled by a red ribbon of activity, and bar encompassing the open kitchen
2-3. Acoustic circles hanging dramatically from the ceiling, and bespoke domes, suspended over intimate booths, creating a series of private spaces within a buzzing public place
4. Private dining area
5. Activity areas for children
6. A new kiosk punching through the restaurant wall, allowing passersby to get food on the go
7. Seating booth along the window

1. 蜿蜒的红色吧台将开放式厨房和起来，使得pizza的制作过程便犹如一场戏剧表演
2、3. 从天花上悬垂下来的吸音圆盘以及悬浮在休息区上方的圆顶结构将整个空间分割开来，形成一系列的单独私密空间
4. 私密就餐区
5. 儿童活动区
6. 餐厅原有的一面墙壁凿开并打造了一个小的售货亭，方便路人购买
7. 沿窗设置的座区

PizzaExpress Plymouth
PizzaExpress普利茅斯分店

Location: Plymouth, UK
Designer: Baynes & Co
Photographer: Baynes & Co
Fitout & construction: Leech Group Services Limited

地点：英国，普利茅斯
设计：贝尼斯设计有限公司
摄影：贝尼斯设计有限公司
施工单位：Leech集团

PizzaExpress like their restaurants to have original features that relate to the locality. Plymouth could be said to be the home of the Royal Navy, so it was a logical starting point to use maritime references for the interior. A huge wavy wall constructed of timber laths forms one side of the restaurant, with infill pieces describing an abstracted wave of water. The timber can also be seen as the hull of a boat. A white painted plimsol line adds an extra layer of complexity to the wall.

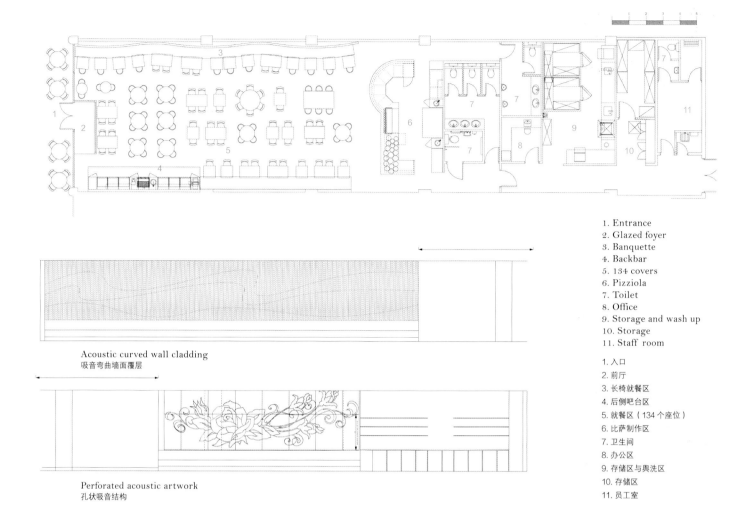

Acoustic curved wall cladding
吸音弯曲墙面覆层

Perforated acoustic artwork
孔状吸音结构

1. Entrance
2. Glazed foyer
3. Banquette
4. Backbar
5. 134 covers
6. Pizziola
7. Toilet
8. Office
9. Storage and wash up
10. Storage
11. Staff room

1. 入口
2. 前厅
3. 长椅就餐区
4. 后侧吧台区
5. 就餐区（134个座位）
6. 比萨制作区
7. 卫生间
8. 办公区
9. 存储区与盥洗区
10. 存储区
11. 员工室

Floor plan 平面图

1-2. Bamboo flooring, marble tables and warm golden upholstery completing the scene
3-4. A huge wavy wall constructed of timber laths forming one side of the restaurant, with infill pieces describing an abstracted wave of water; the timber also being seen as the hull of a boat and a white painted plimsol line adding an extra layer of complexity to the wall
5. A custom acoustic baffle system in suspended foam, with a central feature to contain the air conditioning system
6-7. Rose 'tattooed' onto oak faced paneling in 12mm diameter holes

1、2. 竹子地面、大理石桌子以及金色座椅使空间在视觉上构成一个统一的整体
3、4. 由木板条打造的墙壁蜿蜒伸展着，里面的填充物呈现出水流的波浪造型。整面墙壁犹如一艘小船的外壳，而白色喷漆曲线结构则为其披上了一层外衣，使其看起来更加复杂
5. 专门打造的泡沫声响孔板结构，空调设备容纳其中，同时构成了空间内的特色结构
6、7. 刺绣到橡木板墙壁上的玫瑰图案

The client is always keen to provide good acoustics to encourage conversation. Therefore, the designer designed a custom acoustic baffle system in suspended foam, with a central feature to contain the air conditioning system. On the opposite wall Baynes & Co. commissioned legendary Plymouth tattoo artist Doc Price to create a typical sailors tattoo design. He came up with a rose, which we had 'tattooed' onto oak faced paneling in 12mm diameter holes. The plimsol line paintwork was applied to the panels, which also gives a 'horizon line' effect to the interior. Bamboo flooring, marble tables and warm golden upholstery completes the scene.

PizzaExpress快餐连锁集团乐于在其设计中体现地域特色。普利茅斯可以说是皇家海军的故乡，因此在空间中加入海上元素便成为了这一项目设计的出发点。餐厅一侧，由木板条打造的墙壁蜿蜒伸展着，里面的填充物呈现出水流的波浪造型。整面墙壁犹如一艘小船的外壳，而白色喷漆曲线结构则为其披上了一层外衣，使其看起来更加复杂。

客户一直热衷于营造良好的音响效果，营造舒适的交流氛围。为此，设计师专门打造了一个泡沫声响孔板结构，不仅能够将空调设备容纳其中，更构成了空间内的特色结构。此外，设计师专门委托当地著名的纹身艺术家多克·普里斯（Doc Price）在另外一面墙壁上打造一幅航海主题的画作。他最终选择了玫瑰图案，并将其刺绣到橡木板墙壁上。木板表面部分采用白色喷饰，为空间增添了一道水平线。竹子地面、大理石桌子以及金色座椅使空间在视觉上更加完整。

Jamie's Italian
Jamie's 意大利连锁快餐厅

Jamie's Italian - Authentic affordable Italian Restaurants

The aim of Jamie's Italian was also to create an environment with a 'neighbourhood' feel, inspired by the 'Italian table' where people relax, share, and enjoy each other's company. Jamie's Italian was designed to be accessible and affordable, a place where anyone is welcome and everyone will feel comfortable.

Jamie's Italian restaurant emphasizes the typical hybrid styling it deserves. Tradition inspired, a space filled in with several vintage furniture pieces & tiles, industrial standards ceiling, galvanized metal table tops, colourful pendant lights, a Carrara marble bench, chessboard like floor tiles and beautiful custom designed cupboards and shelves.

Jamie's 意大利连锁快餐厅——价格合理的正宗意式餐厅。

Jamie's 意大利连锁快餐厅的主旨是营造一个类似于"邻里社区中心"的就餐环境，其灵感源于人们休闲、共享的意式餐桌。餐厅内食物价格实惠，欢迎任何一个前来的客人。

Jamie's 意大利连锁快餐厅设计强调多样性，古老的家具、工业化风格天花、抛光金属桌面、彩色吊灯、大理石座椅、棋盘图案地面、定制橱柜等构成空间主要元素。

Jamie's Italian, Westfield
Jamie's意大利风格餐厅韦斯特菲尔德购物中心店

Completion date: 2010
Location: Westfield, London, UK
Designer: Blacksheep
Photographer: Gareth Gardner
Area: 610 m²
Award:
Finalist, Multiple Restaurant Category, Restaurant & Bar Design Awards 2011
Winner, Best Hospitality Environments, Design Week Awards 2011

完成时间：2010
地点：英国，伦敦，韦斯特菲尔德
设计：Blacksheep 设计公司
摄影：加雷特·嘉德纳
面积：610平方米
获奖：2011年度"餐厅与酒吧设计奖餐厅类"入围
　　　2011年度"设计周奖最佳服务环境类"冠军

Bringing Jamie's Character to a Tuscan Kitchen

Blacksheep were approached by Jamie's Italian to formulate a fresh direction for their new restaurant in Westfield - White City. During this time Jamie's Italian currently had 8 sites across the UK. It was up to Blacksheep to work closely with the client to understand the fundamental parts of the brand which made it unique, in order to develop a new direction for Jamie's, which would stand out from previous sites. This was achieved by spending time working in one of their existing restaurants and taking a trip to Italy to fully immerse themselves in Italian culture and discover the passion that first ignited in Jamie.

The site was originally going to be a crèche and as a result had a unique layout which dictated the main body of the design and journey through the space. Subsequently the space was split into three key areas; the Piazza, Market Place and Back Room These areas were all born from the research into Italian dining habits and culture.

The Piazza at the front of the restaurant spills out on to the terrace creating a connection between the internal and external dining area. The market place is a 20m long corridor where the open kitchen is located; opposite a complete retail solution that displays everything from fresh pasta and bread to books and olive oils. This leads through to the back room which has the feeling of a more intimate dining space which you would normally find in the back of an old Italian deli. The journey through from the main entrance to the Rear Room is reminiscent of the scene from *Goodfellas* where the customer is walked through the hustle & bustle of the main kitchen and market into the back, where the walls are lined with hand-made Italian wall panels designed specifically for Jamie's Italian.

Blacksheep designed and developed a feature wall of reclaimed Vesper headlights, all illuminated with LED, that shimmer and glisten through the evening; a nod back to Jamie's humour and history which is offset against beautiful bespoke glass chandeliers and copper fittings. The site is easy to locate due to a Citroen HY Classic van that acts not only as an iconic piece of signage but also as a functioning ice cream van during the summer months. These elements help to define the look of the restaurant, create talking points and make it stand out from the crowd.

The restaurant exterior has a great L-shaped space (approx 100 sqm in total) on Westfield's southern terrace, which offers al fresco dining beneath a canopy. The restaurant's interior is very long (55m from the main door to the door by the antipasti area towards the rear of the restaurant), with a particularly narrow central area, which wasn't ideal for seating. This gave rise to Blacksheep's ideas for the optimum space-plan and for zoned areas, from the main restaurant space (Piazza) to a cen

tral 'Market Place' area, without seating, which made a virtue of the narrow central space and where customers can see views of the kitchens and fresh pasta making or can linger over a bigger retail area than ever before, en route to either the toilets or else to the restaurant's rear dining space, the Back Room, a special, more intimate area with lower ceilings and a slightly more 'bling' treatment.

Interior concept:
Overall, the interior has a pared-down industrial feel. Flooring is a mix of recycled, engineered timber and Black Mountain river slate (used in three different sizes in random patterns).

Walk-through:
When customers arrive, they are directed to a holding bar just inside the main entrance, whilst awaiting a table, but the focal point of the main restaurant area is the dispense bar to the rear of the zoned space and the 'Vespa Wall' in front of the toilet block area. This is made up of a large series of individual headlights which pulsate gently on different settings and can be seen from the outside.

Beyond the machines is the Market Place area, full of noise, bustle, cooking smells and atmosphere.

The final space is the Back Room. A feature wall to the rear of the space features bespoke hand-made timber panels with patinated brass inlay details.

1. The interior boasting a pared-down industrial feel
2. A feature wall to the rear of the space featuring bespoke hand-made timber panels with patinated brass inlay details
3. Chandelier
4. Exterior view
5. Dispense bar to the rear of the zoned space
6. 'Vespa Wall' made up of a large series of individual headlights which pulsate gently on different settings and can be seen from the outside
7. The internal and external dining areas seeming connected together
8-9. Dining areas of different style

1. 室内空间凸显工业化风格
2. 手工制作的木板墙嵌入翠绿色的黄铜装饰，别具特色
3. 吊灯
4. 外观
5. 与就餐区相邻的咖啡吧
6. 卫生间前面的回收汽车前灯背景墙别具特色，从室外也可看到
7. 室内外就餐区看似连通起来
8、9. 不同风格的就餐区

Piazza
1. Main entry
2. Jamie's van
3. Exterior seating
4. Greeter desk
5. Bar

Market place
6. Toilet
7. Disabled toilet
8. 'Market place'
9. Kitchen

The Back Room
10. Kitchen

Preparation
11. Antipasti counter
12. The Back Room
13. Stores
14. Staff
15. Admin

比萨区
1. 主入口
2. 流动贩卖车
3. 室外就餐区
4. 接待区
5. 吧台

展示区
6. 卫生间
7. 残疾人专用卫生间
8. 展示
9. 厨房

就餐区
10. 厨房备餐区
11. 柜台
12. 就餐区
13. 存储区
14. 员工区
15. 办公区

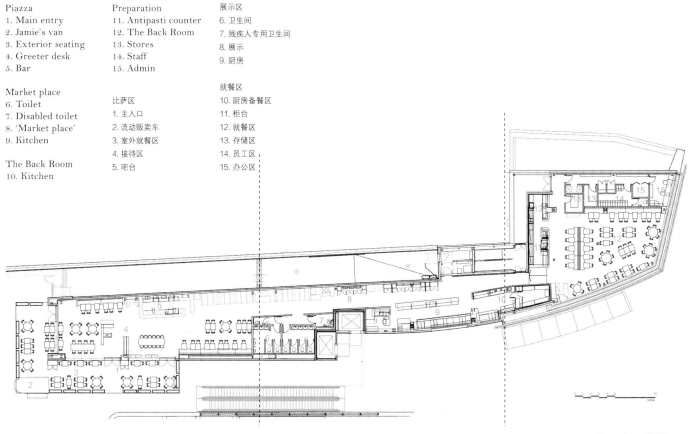

Floor plan 平面图

将Jamie's意大利餐厅的特色引入到托斯卡纳风格的厨房中

Jamie's意大利餐厅委托Blacksheep设计公司为其位于白城韦斯特菲尔德购物中心的新店构思一个全新的理念。当时，Jamie's在英国共有8家分店。设计师同客户进行密切交流以便于了解其品牌的基本构成部分和使其与众不同的元素，旨在打造一个别出心裁的设计。他们曾花费大量的时间对其中一家分店仔细研究并到意大利去感受独特的当地文化，最终实现了他们的目标。

这里曾经是一个幼儿园，独特的格局奠定了主体空间和动线设计的基础。设计师通过对意大利文化和餐饮习惯的调查研究之后，将整个空间分成三个区域：比萨区、展示区和就餐区。

比萨区位于餐厅的前部，一直延展到露台处，将室内外就餐区连结起来。展示区是一条长达20米的走廊空间，开放式厨房便设置在这里。厨房对面摆放着新鲜的面包、图书和橄榄油等。穿过这里，便可达到就餐区，亲切的氛围让人备感舒适。从入口到就餐区的这一段旅程让人不禁想到电影《好家伙》中的场景。客人穿过喧嚣的厨房和展示区，最终达到就餐区，专门定制的手工墙板营造出独特的风格。

设计师专门打造了一面由回收的LED照明汽车前灯构成的特色墙，在傍晚时分熠熠闪烁，反射出Jamie's的特色与历史，同时与定制的玻璃吊灯和黄铜配件相互衬托。此外，餐厅便利的地理位置使得流动贩卖车的引进变得异常容易，不仅作为这里的标识还可以在炎热的夏季售卖冰淇淋。这些设计元素赋予餐厅独特形象的同时，更使其在周围的环境中脱颖而出。

餐厅外面沿韦斯特菲尔德购物中心南侧露台有一个L形的开阔空间，用作室外餐区。餐厅室内长达55米，其中中心区域很窄，不适于就餐。设计师充分利用这一格局特色，将厨房和展示区放在这里，顾客可以在这里驻足欣赏厨房中的场景，之后来到就餐区。

室内设计理念
餐厅室内从总体上来说是简约的工业风格，地面采用回收的木材和三种规格不同的石板铺设。

动线设计
顾客进入餐厅主入口之后，经指引到达前台等待。餐厅的中心区域是自助咖啡吧和就餐区，特色元素则是卫生间前面的回收汽车前灯背景墙，从室外也能看到。

厨房展示区是一个喧嚣的场所，这里充斥着噪音和食物的味道。

最后的就餐区，手工制作的木板墙嵌入翠绿色的黄铜装饰，别具特色。

Mangiare

Mangiare 连锁快餐厅

The Italian Revolution - Good food fast

The particular inspiration behind Mangiare is the pizza-al-taglio sold throughout Italy as a simple, delicious and inexpensive lunch. Pizza-al-taglio translates as cut pizza, as each customer's portion is cut from a large rectangular tray.

Mangiare has refined this original generic Italian formula into a high quality Italian fast food concept for the UK, that meets the growing demand for fast, convenient, premium quality working lunches – and the desire for hot wholesome food.

In restaurant space design, colour of dark grey, exposed ceiling, decorative screen of oak and grey/yellow Formica incorporating menu boards and signage, simple lines are commonly used.

意式食物的改革——高品质快餐

Mangiare 连锁快餐的灵感源于"切片比萨饼"（在意大利非常流行的食物，通常作为简单美味而又价格实惠的午餐）。顾客可以根据自己的需求从一整块比萨中切取一部分购买。

Mangiare 连锁快餐打破了传统的形式，成为高品质的意式快餐，满足了顾客对于食物供应速度、便利性和质量的要求。

餐厅空间设计中，深灰色、裸露天花、镶嵌在橡木构成的装饰屏风内的菜单板和标识板、简约的线条灯元素成为 Mangiare 的特色，并被广泛应用。

Mangiare Spitalfields

Mangiare快餐斯皮特菲兹店

Completion date: 2010
Location: London, UK
Designer: Jonathan Clark Architects
Photographer: Jonathan Clark Architects
Area: 150m²

完成时间：2010
地点：英国，伦敦
设计：乔纳森·克拉克建筑事务所
摄影：乔纳森·克拉克建筑事务所
面积：150平方米

A new fit-out for Mangiare is currently on site - within the recently completed Nido development. This was the complete fit-out of a basic 'shell & core' retail space in Spitalfields, London E1. Brief was to create a modern streamlined environment with strong graphic elements on a very tight budget. As an Italian fast food & takeaway, all surfaces had to be hardwearing with easy maintenance. Internal area including new kitchen is 150m² (1,600ft²) and maximum seating covers of 60 were achieved.

The interior is pretty much a distillation of the features produced for the previous 'Mangiare'. Very simple lines with the simple graphic yellow and dark grey Formica cladding contrasted with white Carrara marble counter tops, bare blockwork walls painted cream and oversized black & white photographic posters.

A decorative screen of oak and grey/yellow Formica incorporating sliding blackboard menus runs full width behind the counters – helping to exaggerate the size of the space as well as provide a visually stimulating backdrop for queuing customers.

There had generous ceiling heights and the designers decided to keep new services exposed to maintain this as well as save the costs of providing a suspended ceiling. The kitchen and staff serving areas were raised in order to overcome some awkward drainage positions and create the maximum width of counters for pizza and pasta display and tills etc.

They were also responsible for the design and implementation of the signage and graphics including the shopfront.

1

1. Shop front
2. A decorative screen of oak and grey/yellow Formica incorporating sliding blackboard menus running full width behind the counters
3. Very simple lines, simple graphic yellow and dark grey Formica cladding and ceiling of various heights
4. Oversized black & white photographic posters adding a unique feeling to the space

1. 店面
2. 菜单板和标识板镶嵌在一个橡木构成的装饰屏风内,顺着餐台后面摆放
3. 简约的线条、黄色及深灰色的板材覆面结构以及高低不一的天花
4. 大幅黑白海报给空间带来独特的感觉

Mangiare快餐斯皮特菲兹店位于Nido开发区内,这一项目包括室内外空间整体设计。客户要求利用有限的预算营造一个具备现代风格、平面元素丰富的流线型空间。餐厅内提供外带服务,要求所有结构的表面必须符合耐磨及易清洗的要求。室内包括一个新增的厨房,可容纳60人就餐。

室内空间集合了Mangiare的各种特色元素,简约的线条和深灰色的板材覆面结构与白色卡卡拉大理石台面、裸露的米色砖石墙壁和黑白的大幅海报形成鲜明的对比。

菜单板和标识板镶嵌在一个橡木构成的装饰屏风内,顺着餐台摆放,既在视觉上增大了空间面积,同时也为排队的顾客提供了一个美好的背景。镜子被大量运用,进一步增大了空间开阔感。

高度不一的天花被保留下来,从而节约了打造假天花的预算;厨房和员工区被提升,以越过排水点,同时增大了餐台的宽度。

此外,设计师还负责标识、菜单以及店面等平面设计工作。

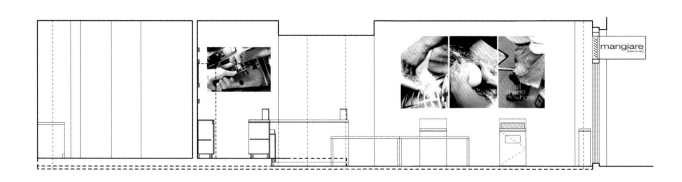

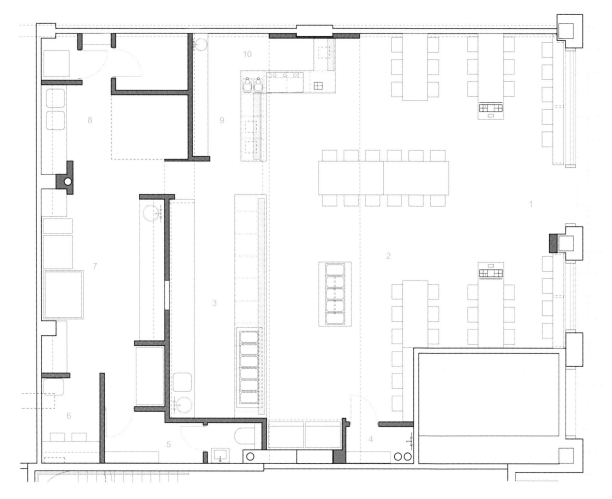

1. Entrance
2. Main area
3. Main counter
4. Cleaners c/b
5. Staff
6. Office
7. Kitchen
8. Wash-up
9. Till point
10. Coffee

1. 入口
2. 主就餐区
3. 柜台
4. 清洁用品区
5. 员工室
6. 办公区
7. 厨房
8. 盥洗区
9. 其他物品售卖区
10. 咖啡区

Floor plan 平面图

Mangiare
Mangiare快餐伦敦店

Completion date: 2008
Location: London, UK
Designer: Jonathan Clark Architects
Photographer: Jonathan Clark Architects
Area: 120m²

完成时间：2008
地点：英国，伦敦
设计：乔纳森·克拉克建筑事务所
摄影：乔纳森·克拉克建筑事务所
面积：120平方米

Mangiare is an upmarket Italian fast food outlet on London Wall in the city. The premises were previously a branch of a well known café chain and were totally stripped out to reveal 3.6m high ceilings with some original Victorian cornices and dado rails at high level.

Two internal structural columns were retained and a 5.5m long Carrara marble clad communal table with purpose designed/made benches straddles these. Dark grey and bright yellow Formica and oak was used throughout to clad various internal elements. New servery counters with wany edged oak boards were inserted separating the public areas from the new kitchen with its imported Italian oven.

The existing ceiling above the previous suspended ceiling tiles was left and painted dark grey to 'disappear'. Exposed mechanical services (ventilation, heating and cooling) were installed at high level and are carefully coordinated with the new lighting installations.

A decorative screen of oak and grey/yellow Formica incorporating menu boards and signage runs from front to back behind the counters – helping to exaggerate the size of the space as well as provide a visually stimulating backdrop for queuing customers. Areas of mirror are also utilized to give a greater sense of volume.

The designers were also responsible for the design and implementation of the signage, menus and graphics including the shopfront.

1

Mangiare是一家高档意式连锁快餐厅,位于伦敦墙(London Wall,伦敦市区的一条大街)旁。这里曾经是一家连锁咖啡厅,经彻底改造之后,建筑原有的3.6米高天花被彻底裸露出来,维多利亚风格的檐口和护墙围栏别具特色。

室内两根构造柱被保留下来,一张长5.5米的卡拉拉大理石覆面长桌以及特殊定制的座椅横跨其间。深灰色和亮黄色的橡木材质被广泛应用,装饰不同结构的表面;橡木板打造的备餐台将公共就餐区与新建的厨房分隔开来。

天花上的瓷砖被刷成深灰色,在视觉上营造出空旷感。通风、供热及制冷系统直接裸露在外,并与新安装的灯饰融合在一起。菜单板和标识板镶嵌在一个橡木构成的装饰屏风内,顺着餐台摆放,既在视觉上增大了空间面积,同时也为排队的顾客提供了一个美好的背景。镜子被大量运用,进一步增大了空间开阔感。

此外,设计师还负责标识、菜单以及店面等平面设计工作。

1. Communal dining table and salad bar
2. Counters with oak panels cladding on the front
3. Menu boards incorporated in a decorative screen of oak
4-5. Dining tale of different styles

1. 就餐长桌和沙拉吧台
2. 柜台正面采用橡木板饰面
3. 菜单板镶嵌在橡木装饰屏风内
4、5. 不同风格的就餐桌

1. Counter
2. Table
3. Communal table
4 Coffee Tills
5 . Fridge
6. Salad bar
7. Pizza counter
8. Prep kitchen
9. Pizza oven

1. 柜台
2. 桌子
3. 就餐长桌
4. 咖啡区
5. 冰箱
6. 沙拉吧
7. 比萨柜台
8. 备餐厨房
9. 比萨烤箱

Floor plan 平面图

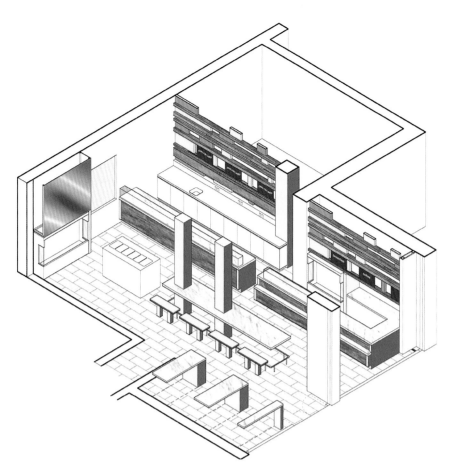

Client's Comments:
'I have worked with JCA on a number of different projects and have always been very impressed by them.
They always took the time to understand what I was trying to achieve and on each occasion provided me with the best solution and delivered it on time and within budget.' Carlo Ventisei, Mangiare

来自于客户的评价：
"我曾经和乔纳森·克拉克建筑事务所合作过多次，每次都会被他们感动到。他们经常花掉很长时间去了解我想要达到的目标，而且每一次都会在预算和较短的时间内帮我们找到最合适的方式。"卡洛·文蒂塞

cocos

Cocos 快餐厅

The philosophy behind the bok and cocos bistros is all about enticing people in busy locations to take a quick break in their everyday life to enjoy high-quality food. The design for cocos is based on the contrast between the bustle of the outlet centre and people's desire for rest; allowing guests to find peace and quiet is given prime focus here. The understated design means that the space appears to be homogeneous, and the eye can come to rest.

Cocos 快餐厅的设计理念是为顾客营造一个小憩的空间,同时提供高品质食品。设计强调对比——餐厅内的繁忙景象与顾客渴望休息的神情交织,平静宁和的空间氛围便成为了设计的重中之重。简约而低调的设计使得空间备显宁静。

cocos Parndorf

cocos帕斯多夫店

Completion date: 2011
Location: Parndorf, Austria
Designer: Heyroth & Kürbitz freie Architekten BDA
Photographer: Maks Richter, Stuttgart
Area: 176m², 58 seats
Client: bok & cocos GmbH

完成时间：2011
地点：奥地利，帕斯多夫
设计：Heyroth & Kürbitz建筑设计公司
摄影：麦克斯·理奇（斯图加特）
面积：176平方米（58个座位）
客户：bok&cocos有限公司

The design: a self-service, 17 meters-long with a dramatic gesture.

cocos Parndorf invites you to take a seat. Anyone entering the Asian bistro is immediately attracted to the 17 meters-long counter which forms the heart of the restaurant in the shape of an orangey-red meander. The visitor in a hurry can help himself to exactly what he wants at the buffet, whilst the aficionado can take a look at what is going on in the kitchen – screened off by a glass wall. If you have created your own menu, you will have time to sit peacefully and enjoy your meal. Guests can turn their backs on the kitchen glowing like a flame in the middle of the restaurant and seat themselves in the dark green tones of the seating area, in order to enjoy their well-earned meal. The bold curve of the counter finds a mellow echo here. White, undulating patterns break up the lush green colour of the walls, again finding an echo on the ceiling in the three-dimensional form of curved panels, which repeat the meandering shape of the counter.

The challenge: Cuisine on a grand scale.

The 'Design Outlet Centre' in Parndorf has a high footfall. It was the aim to find a spatial solution which provides room for the stream of visitors and also permits optimal working conditions.

Floor plan 平面图

1. Service counter　1. 服务柜台
2. Open kitchen　2. 开放式厨房
3. Dining area　3. 就餐区
4. Toilet　4. 卫生间
5. Staff room　5. 员工室

1-2. The red lighting fixtures dangling from the green ceiling seeming like blooming flowers
3-4. White, undulating patterns breaking up the lush green colour of the walls
5. The 17 meters-long service counter standing in the centre with a dramatic gesture
6-8. The ceiling in the three-dimensional form of curved panels corresponding with the undulating counter and white lines on the wall

1、2. 红色的灯饰从绿色的天花上悬垂下来，犹如盛开的花朵一般
3、4. 白色蜿蜒的图案打破了墙壁上连续的深绿色调
5. 17米长的服务柜台以一种独特的姿态矗立在餐厅中央
6~8. 天花上立体的曲线板结构与蜿蜒的柜台以及连续的白线条交相呼应

设计: 17米长的自助柜台以一种独特的姿态矗立在餐厅内。

蜿蜒的形态以及橘红的色调使其自然而然地成为整个空间的中心，并欢迎着顾客的到来。匆匆忙忙的顾客可以在自助柜台上选取自己想要的东西，而那些时间充裕的客人则可以看看玻璃后面厨房中正在加工的食品，点好之后静静地在这里等待。他们可以坐在背对着厨房的深绿色调的休息区内，细细地品味可口的食物。柜台的弯曲造型被映射到墙壁上，留下一道道柔和的曲线造型。白色蜿蜒的图案打破了墙壁上连续的深绿色调，并与天花上立体的曲线板结构交相呼应，进一步强调了柜台的蜿蜒形态。

挑战：营造一个宽敞的就餐环境。

设计的目标即为找到一种方式，以便于为顾客提供宽敞的就餐空间，同时符合最佳工作流程要求。

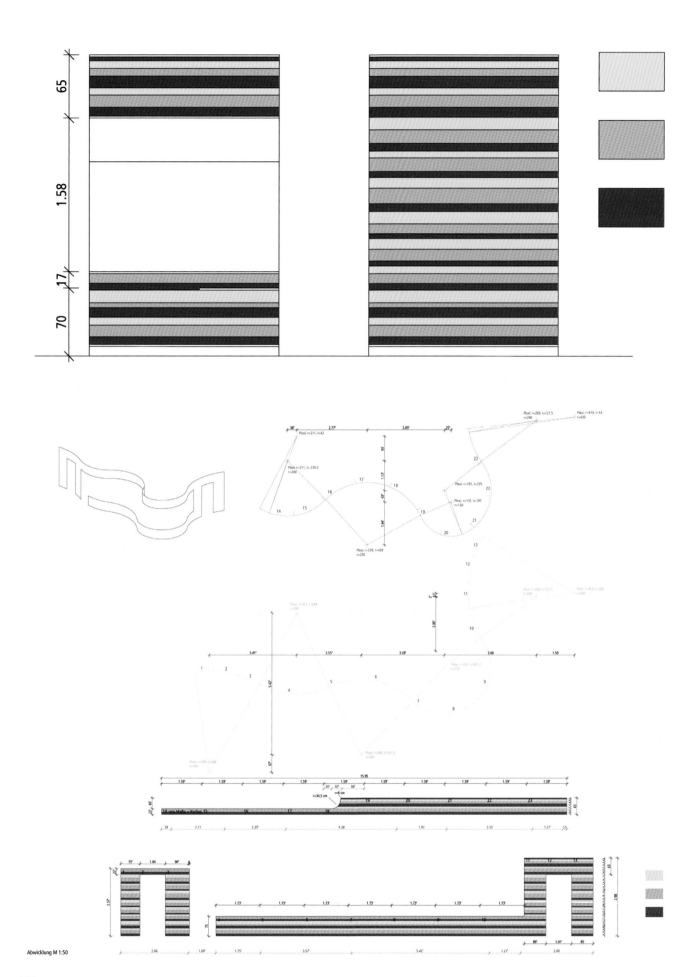

Abwicklung M 1:50

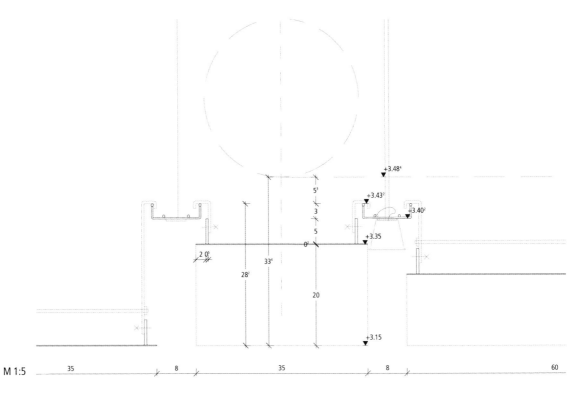

M 1:5

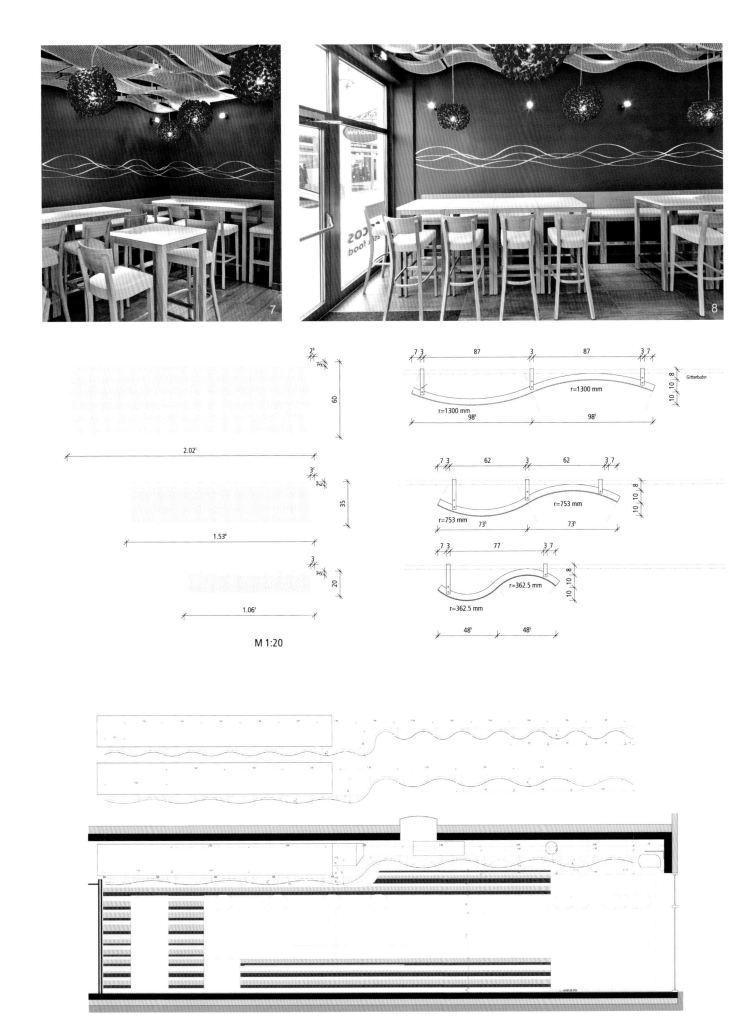

cocos Passau
cocos帕绍店

Completion date: 2008
Location: Passau, Germany
Designer: Heyroth & Kürbitz freie Architekten BDA
Photographer: Christian Haasz, Tittling
Area: 68.8m², 36 seats
Client: bok & cocos GmbH

完成时间：2008
地点：德国，帕绍
设计：Heyroth & Kürbitz建筑设计公司
摄影：克里斯蒂安·哈茨
面积：68.8平方米（36个座位）
客户：bok & cocos 有限公司

The design: dining beneath the willows

Like rays of sunshine, light is filtered through the filigree leaf canopy made of plexiglass, conjuring a play of reflections and shadows over walls and floor. Extending wall to wall, the translucent foliage has the effect of dissolving the ceiling.

The soft green canopy of the restaurant draws the eye. The warm brown tones of the furniture, flooring and bar provide a gentle, understated backdrop to the animated patterns of light. The earthy quality of the other fittings forms a contrast to the weightlessness of the ceiling installation. To augment the garden ambiance, small porcelain lamps, shaped like Chinese lanterns, bathe each individual table with soft lighting.

The challenge: nature transformed

In order to integrate the theme of nature within architecture, without relying on simple imitation is no easy task, yet we successfully achieved this at cocos in the Stadtgalerie in Passau. Kerstin Heyroth: 'Nature is to be found in every kind of architecture. It is our inspiration, demanding to be reworked in a contemporary manner.'

1

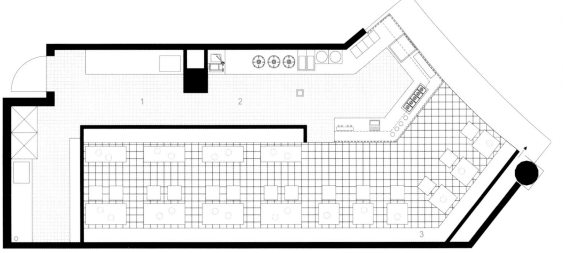

1. Preparation (16.0 m²)
2. Kitchen (16.4 m²)
3. Dining (38.4 m²)

1. 备餐区（16平方米）
2. 厨房（16.4平方米）
3. 就餐区（38.4平方米）

Floor plan 平面图

1. Shopfront
2. Plexiglass willow pattern in green colour creating a fresh and natural space
3. Logo, menu board and counter
4-5. Small porcelain lamps, shaped like Chinese lanterns

1. 店面
2. 有机玻璃打造的绿色柳条图案营造了一个清新而自然的空间
3. 标牌、菜单板与柜台
4、5. 娇小的陶瓷灯犹如中式灯笼一般悬垂下来

1. Mirror full height, titled by 1° up	1. 镜子 等高,向上倾斜1度
2. Window pattern plexiglass Satinice Kiwi 6mm sheet 196 X 116 laser-cut mounted by nylon coil + hook	2. 窗户样式 磨砂板6毫米(具有光漫射功能) 196x116 激光切割板 尼龙线圈和挂钩
3. Wall	3. 墙壁
4. Matte white paint	4. 白色亚光漆
5. Bench	5. 长凳
6. Installations	6. 装置
7. Dish return	7. 餐具回收处
8. Menu board	8. 菜单板
9. Glass counter	9. 玻璃柜台
10. Counter from HPL digital print	10. 柜台 数码印花板
11. Logo	11. 标识
12. Walk-through	12. 过道区
13. Counter plinth V2A	13. 柜台底座

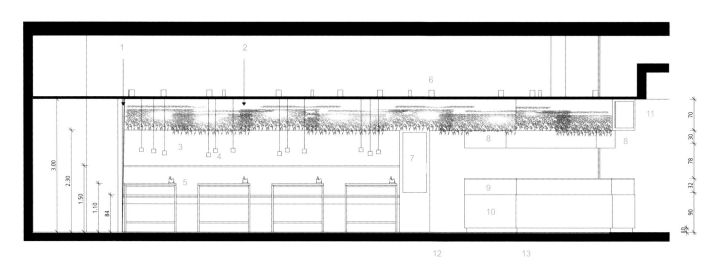

设计：在柳树下美餐

灯光犹如太阳光线一般穿过由有机玻璃材质打造的树叶图案顶棚，反射在墙壁和地面上，带来了浓郁的趣味性。半透明的树叶图案从一面墙壁延伸到另一面墙壁上，似乎将天花分隔开来。

淡雅的绿色顶棚格外吸引眼球，而棕色调的家具、地面及吧台则为动感十足的图案构成了一个柔和而低调的背景。轻盈的天花装置与其他装饰形成鲜明的对比，娇小的陶瓷灯犹如中式灯笼一般悬垂下来，使得桌子周围沉浸入柔和的光线内，恰似在花园中一样。

挑战：自然理念的融入将自然融入到建筑中本身并不是一件容易的事情，然而在这基础上摒弃简单地模仿则更是难上加难。这一餐厅的设计则成功地实现了这项艰难的任务。

项目设计师克斯廷·海洛斯说："自然元素在各种建筑中都可以找到，这就是我们的灵感源泉，同时我们更希望以一种现代化的方式实现自然与建筑的融合。"

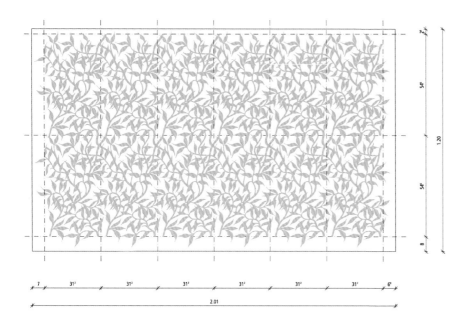

1. Spotlights as sunspots
2. Wall primed Q3
3. Matte wall paint, white
4. Bench
5. Spotlights back lighting willow pattern
6. Plexiglass willow pattern
   -mounted in 4 levels with distance of 5 cm each
   -Mounting:
   -anchor with screw hook in plasterboard ceiling
   -chain
   -S-hook
7. Screwed joint
8. Pendants Eiden
9. Bistro table 45 X 120
   Bistro table 60 X 120
   Bistro table 60 X 60
10. Wall tiles
    -front area, full height
11. Overhead guest room + soffit:
    -matte wall paint
    -kitchen: HPL white
12. Backlit menu board
    -LED-lights, warm white
    -kitchen side: HPL white
    -spotlights for counter lighting
13. Glass counter
    -glued, no fittings
14. Besteckhalter
15. Worktop surface
    -creme white
16. Counter front
    -HPL digital print
17. Plinth V2A-sheet
18. Tiles restaurant 20 X 20
19. Tiles kitchen 10 X 10
20. Wall tiles
    -back area, h=2.20 m

1. 聚光灯
2. 墙壁
3. 白色亚光墙面
4. 长凳
5. 柳条图案采用聚光灯背光照明
6. 有机玻璃柳条图案
   —四层悬挂，每层距离5厘米
   —连接方式：
   —石膏天花上的螺旋钩
   —链条
   —S形挂钩
7. 螺纹接头
8. 吊灯
9. 就餐区桌子规格
   45x120
   60x120
   60x60
10. 瓷砖墙面
    —餐厅前部
11. 拱腹
    —亚光墙面
    —厨房：白色板材
12. 背光照明菜单板
    —LED灯，温馨的白色
    —厨房四周：白色板材
    —柜台聚光灯照明
13. 玻璃柜台
    —黏合，无需采用其他结构固定
14. 盛放刀叉的容器
15. 工作台表面
    —乳白色
16. 柜台正面
    —数码印花板
17. 底座
18. 就餐区瓷砖 20x20
19. 厨房瓷砖 10x10
20. 墙面瓷砖
    —后部，高2.2米

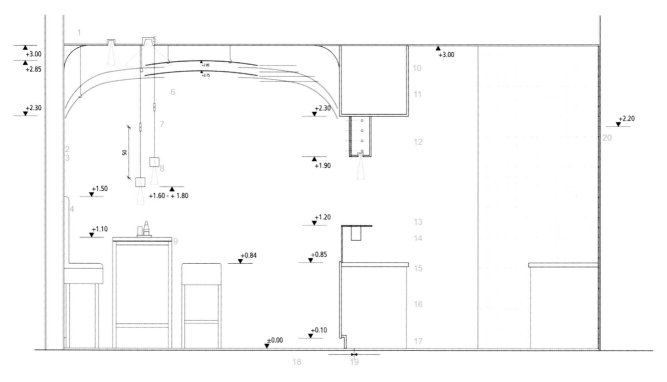

sosushi

Sosushi

创意寿司

It is read as Sosushi but it translates as 'creative sushi', a new way of conceiving sushi that achieves an entirely new balance between tradition and innovation, between the culinary culture of a people and the wisely orchestrated contamination of Italy's excellent cuisine. It is a brand that has been capable of attaching its corporate vision to a name and a colour. Sosushi is magenta and magenta is strength, energy, imperiousness. A colour that has succeeded in binding itself to the embryonic vision of the brand, and it has become a symbol capable of diffusing the image of a brand which is indeed cuisine, but also design, pop art, creativity, woman, Japanese culture and the certainty of a strong 'Made in Italy'.

Sosushi 可译为"创意寿司",以新的方式诠释寿司这一食物。其在传统和创新之间、日式饮食文化和意大利菜式之间取得了完美的平衡。Sosushi 已成为一个品牌,将企业形象与名字和一种色彩紧密地联系起来。Sosushi 是品红色的,品红色是力量、活力的象征。Sosushi 成功地将品红色融入到品牌形象中,使其不仅仅是一种食物,更代表着设计、流行艺术、创意、日式文化与意大利韵味。

Sosushi Rho

Sosushi寿司店罗镇分店

Completion date: 2009
Location: Rho, Milan, Italy
Designer: Luca Bertacchi & Sara Bergami
Photographer: Josep Pagans

完成时间：2009
地点：意大利，米兰，罗镇
设计：卢卡·贝尔塔基&萨拉·贝加米设计公司
摄影：约瑟普·帕甘斯

Sosushi Rho is perhaps the synthesis of functional and material research done for the restaurant chain Sosushi. The restaurant is located close to the centre of the homonymous city, near the Fiera Rho-Pero, and for this reason they've tried to create a contemporary, dynamic and an unconventional space able to charm and intrigue even the most sophisticated customers.

The place is a pure single volume, a cube facing the outside with two large windows that provide maximum transparency and the maximum ratio between the inside and the outside. The idea was to deny any physical limits using optical three dimensional surfaces, tables, lamps, desk and decoration to summon the circle staying close to the aesthetics of the sushi itself. There are no more edges, there is no boundary between the floor and the walls, but a new dynamism and a new dimension of dynamic perception of interior space.

The spot light punctuates the rhythm of the ceiling illuminating a cascade of 1,000 origami hanging in the air. The designers wanted to create a forest of origami that according to Japanese tradition are sign of prosperity and happiness and at the same time create a scene, an installation that could also be an opportunity to attract new artists and designers by giving readings always new.

The furniture, custom made, are designed to coordinate with the design of the walls and floor drawing curves and sinusoidal patterns. The counter of the Japanese chef is always in view to ensure maximum transparency in the production of sushi and living him the role of the real protagonist of the scene.

An artisanal laboratory with a metropolitan and youthful spirit; a concept that involves a fully visible kitchen and the presence of a chef that prepares freshly-made dishes; this is the standard formula complemented by take-out, catering and delivery services.

1. Entrance
2. Dressing
3. Disabled toilet
4. Toilet
5. Deposit
6. Fridge
7. Freezer

1. 入口
2. 更衣室
3. 残疾人专用卫生间
4. 卫生间
5. 寄存处
6. 冰箱
7. 冷冻区

Floor plan 平面图

1. Logo
2. The spot light punctuating the rhythm of the ceiling illuminating a cascade of 1,000 origami hanging in the air
3. Two large windows providing maximum transparency and the maximum ratio between the inside and the outside
4. White table, chairs and bar
5-6. The design of the walls and floor drawing curves and sinusoidal patterns coordinated each other

1. 标识
2. 聚光灯打破了天花的节奏，1000只折纸图案犹如小瀑布一般，在灯光的照耀下，格外引人注目
3. 两扇大窗户确保了空间的通透性，同时将室内外连通起来
4. 白色的桌椅与吧台
5、6. 墙面与地面上的曲线图案相互呼应

Sosushi寿司店罗镇分店的设计或许是为其连锁集团关于"功能与材质的结合"的理念所做的一个实验性项目。餐厅选址在米兰市中心Rho-Pero展览馆附近，因此其目标是打造一个现代风格十足又极具活力的空间，以去吸引那些最为独特的顾客。

从外观看上去，餐厅呈现出简约的立方体造型，两扇大窗户确保了空间的通透性，同时将室内外连通起来。设计要求摒弃空间之间的物理界限，立体感十足的表面装饰让人不禁联想到寿司本身的美感。空间似乎没有边缘，地面和墙面之间没有任何界限，一个动态十足的空间完美呈现出来。

聚光灯打破了天花的节奏，1000只折纸图案犹如小瀑布一般，在灯光的照耀下，格外引人注目。客户意图打造一个折纸森林，折射日本的传统文化（折纸在日本是繁荣和幸福的象征），并营造一种场景，吸引更多的设计师和艺术家前来。

定做的家具与墙面及地面上的曲线图案相互呼应。大厨制作寿司的场景透过柜台后面的玻璃清晰可见，让他成为这里当之无愧的主角。

这家餐厅可以被视作一个带有国际化风格和年轻精神的手工实验室，开放式厨房和大厨制作食物场景的可见以及提供外带、就餐和配送等服务可以说是快餐厅的标准模式。

Sosushi Train Turin
Sosushi寿司店都灵火车站分店

Completion date: 2010
Location: Turin, Italy
Designer: Luca Bertacchi & Sara Bergami
Photographer: Luca Bertacchi

完成时间：2010
地点：意大利，都灵
设计：卢卡·贝尔塔基&萨拉·贝加米设计公司
摄影：卢卡·贝尔塔基

It is inspired by the idea of a journey, built with the concept of movement, and it guarantees a service where flexibility and mobility are at the forefront. It has kaiten and it is perhaps the most fascinating format of the range. As with the others, there is take out, delivery and naturally the possibility to eat there and then.

The cherry tree for Japanese culture is SAKURA, a metaphor for the transience of life: relentless journey that starts from the sap, leading to a flower and gives life to the fruit; a lush and blooming fair, fleeting but that brings with it a vivid and penetrating beauty. In Japan the cherry blossom has a place of honour and it even becomes the national flower. In this flourishing is the symbol of transience, and it is from this that the idea for this restaurant was born.

The restaurant is located in one of the most important station in the country: Turin. The design has been developed taking into account the parallel between the cherry tree and its transience and the trip, main character of the location. The sap that flows through a tree and that travelling from one place to another is capable to blossoming flowers giving life to lush fruit as a vivid transience. However the proximity to the TRAVELLER, who habitually try frequents places of travelling, passage and transit passengers only, nomads running along preferential pathways ready to lead them to the longed-for destination ... the desired destination: under a cherry tree.

Standing under a cherry tree in spring you can look up and see pink flowers everywhere, and summer fruits, berries everywhere. 'We would liked to see the travellers enter the room and look up at the ceiling and feel as if they were under a cherry tree.'

1

1-2. Logos on the wall
3-4. Lights seeming floating on the distinctive ceiling
5-7. Patterns of cherry tree on the wall to interpret the concept of the design

1、2.墙面上的标识
3、4.灯饰 似乎悬浮在独特的天花上
5~7.墙上的樱花树图案诠释出设计的主题

这一设计源于"一次旅行",以运动为主体理念,保证为顾客营造集灵活性和流线型于一身的完美体验。回转寿司的形式更令人期待,这里更提供外带、送货等服务。

樱树在日本文化中被称作"Sakura",代表着生命的轮回,从发芽到开花再到结果,便完成了一次生命的旅程。开花结果的过程似乎转瞬即逝,但却留下了不可磨灭的美。在日本,樱花是荣誉的象征,并因此被作为国花。繁茂地生长就是一段旅程,这一餐厅的设计灵感正是源自这里。

餐厅位于都灵火车站,其设计理念源自樱树的成长之旅,而旅程更是与车站紧密联系在一起。从一个地方到另一个地方的旅行过程如同樱树从开花到结果,虽然短暂,但却能够为生命增添内容。对于旅行者们来说,他们可以选择自己钟爱的方式和路线,最终到达渴望已久的目的地——樱树下。

春天,站在樱树下,抬头看见粉色的花瓣;夏天,则是红色果实。"我们希望营造的就是这样一个场景,顾客走进餐厅,抬头便可看见天花上的樱树,感觉自己如同置身在樱树下。"

1. Restaurant
2. Kitchen
3. Reception

1. 就餐区
2. 厨房
3. 接待区

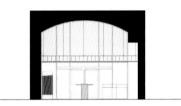
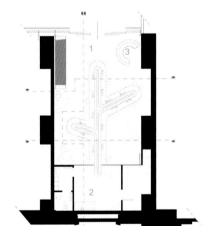
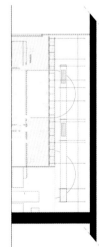

Floor plan 平面图

Sosushi Sassari
Sosushi寿司萨萨里店

Completion date: 2009
Location: Sassari, Italy
Designer: Luca Bertacchi & Sara Bergami
Collaborator: Carolina Semeghini
Photographer: Josep Pagans

完成时间：2009
地点：意大利，萨萨里
设计：卢卡·贝尔塔基&萨拉·贝加米设计公司
合作设计：卡罗丽娜·塞米吉尼
摄影：约瑟普·帕甘斯

Sosushi Sassari was previously a garage with a clear industrial taste, a long corridor, a ramp and a large space but not too high, was the first time compared with a similar size. The client wanted that there were no limits, and barriers. As caught by a sudden gust of wind, the inside becomes the outside and in a moment one could be transported to the upper floor.

The designers imagined a long segmented wall, like an open origami, strongly tied to tradition, that could embrace all the space inside and bring the customer entrance to the hall, through its shadows and its fragments to the counter where the Japanese chef will be the protagonist.

A light wall suspended between two long strip of light that besides defining the importance of lightness it gives to customer a new perception. The floor is neutral with a warm grey colour that reminds customers of what lies outside the space.

They wanted to play on contrasts, however, by recalling some elements typical of this restaurant chain. A delicately coloured origami wall, a grey floor, a wall along with a contemporary stylized design imprinted in the material.

The light holds together the individual elements putting them in the right order. At the entrance they put an iconographic chandelier that welcomes customers and shortly introduces them to a forest of lighten balls of various forms and intensity that coexist with the wall origami suspended between two strips of light, lower and higher. In the upper room, then the customers will find bright bubbles coexist with more contemporary lamps positioned above the counter and they will find an intimate corner with soft light where a large round table brings them back home in the living memory and in the pleasure of eating together.

1. Reception　　1. 接待区
2. Entrance　　 2. 入口
3. Bar/restaurant 3. 吧台/就餐区
4. Storage　　　4. 存储区
5. Kitchen　　　5. 厨房

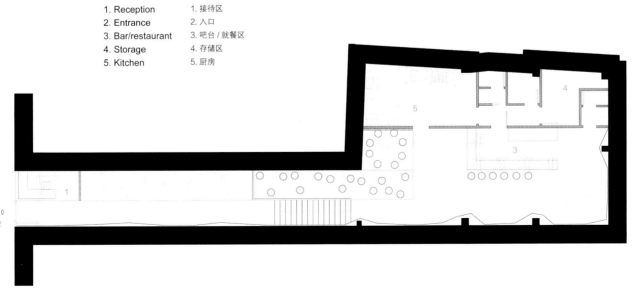

Floor plan　平面图

1. A long segmented wall, like an open origami, strongly tied to tradition
2-3. Decorations on the wall reflecting features of Sosushi
4. Bright bubbles coexisting with more contemporary lamps positioned above the counter
5. Ceiling
6. Staircase
7. The floor neutral with a warm grey colour reminding customers of what lies outside the space
8-9. Dining area

1. 一面分段式墙壁，犹如折纸一般，突出寿司发源地的文化
2、3. 墙面上的装饰凸显Sosushi的特色
4. 明亮的灯泡和现代风格的灯饰组合在一起
5. 天花
6. 楼梯
7. 地面采用灰色调装饰，让人情不自禁地想到室外的空间
8、9. 就餐区

这里之前曾是一个工业气息十足的车库，由走廊、坡道和一个开阔但不高的空间构成。对于设计师来说，他们是第一次以这样格局的空间为工作对象。客户要求打破室内外及不同空间之间的界限，比如，一阵风吹过，室内瞬间便成为了室外，或者短暂的时间内，顾客就可以从一楼到达二楼。

为满足客户的要求，设计师打造了一面分段式墙壁，犹如折纸一般，将室内的所有空间围合起来，同时又突出了寿司发源地的传统文化。顾客走进大厅，随着光影的引导，便可到达柜台，欣赏主角——日本大厨的"表演。"

一面嵌灯墙介于两排灯饰之间，不仅突出了光线的重要性，更给顾客带来全新的视觉享受。地面采用灰色调装饰，让人情不自禁地想到室外的空间。设计师乐于运用对比，Sosushi连锁餐厅的特色元素与精心打造的彩色墙壁、灰色地面等形成鲜明的对比。

光将单独的元素集合在一起，并将它们放到适当的位置上。入口处，一盏吊灯欢迎着顾客的光临，之后犹如进入一个光球的森林，不同样式和强度的灯饰从天花上悬垂下来。楼上，明亮的灯泡和现代风格的灯饰组合在一起。在这里，顾客可以找到一个光线柔和的角落，围着一张大圆桌坐下来就餐，如同回到家中一般美好。

Costa Coffee
咖世家

The concept of Costa Coffee is to deliver to a younger, trendier, more cosmopolitan audience with the flexibility to change every part of the brand concept except the logo.

咖世家（Costa Coffee）的理念即为"多样性"，除品牌标识之外，其他部分均可灵活变化，以传递出年轻、时尚的国家化特征。

Costa Coffee, Great Portland Street

咖世家大波特兰街店

Completion Date: July 2010
Location: London, UK
Designer: Stiff + Trevillion
Photographer: Kilian O'Sulivan
Area: 204 m²
71 indoor seats, 16 outdoor seats and 10 standings
Award: Cafe or Fast Food - Restaurant & Bar Awards 2011
Best Interior – BCSC Gold Award 2011

完成时间：2010.7
地点：英国，伦敦
设计：Stiff & Trevillion事务所
摄影：基利安·欧苏力文
面积：204平方米（71个室内座位，16个室外座位，10个站位）
获奖：2011年度餐厅酒吧设计奖
　　　2011年度BCSC最佳室内设计金奖

The first 'metropolitan' Costa opened in summer 2010 on a corner site in central London's Great Portland Street.

Following a competition in early 2010, architects and designers, Stiff & Trevillion were appointed to design and implement the new concept, catering specifically for the urban customer. Costa had been performing very well but research revealed an opportunity for the brand to better meet the requirements of customers in large urban areas, who are surrounded by an eclectic choice.

The aim with the new metropolitan stores was to strengthen the message around 'coffee is king', making this a key point of the interiors and customer journey. An individual look and feel for each of the sites using existing elements such as exposed brick walls and ceilings are coupled with concrete flooring and statement lighting to give a real utility feel. The powerful in-store graphics, photography and artwork 'hero-ing' the coffee bean and elements of its by-products have been beautifully selected, crafted and illustrated by Helen Senior Associates, who worked closely with Stiff & Trevillion on realizing the project.

Costa has two types of customer; the 'express' customer comes in for their takeaway coffee and bites, whilst the 'relax' customer dwells in store refreshing themselves whilst reading a book, working on their laptop or meeting friends and colleagues. The award winning Great Portland Street store is designed to be a hybrid experience for both of these customer types and as such has two distinct zones.

The recent radical re-launch of Costa coffee's urban café has proved very successful for the chain, resulting in a move to expand the concept. The new 'metropolitan' Costa, as it is known, currently has three sites in London with another 10 or so earmarked for the coming year.

1. Entrance
2. Counter
3. Express dining
4. Relax dining
5. Toilet

1. 入口
2. 柜台
3. 快餐就餐区
4. 休闲就餐区
5. 卫生间

Floor plan 平面图

1. Shopfront
2-3. Exposed brick walls and ceilings coupled with concrete flooring and statement lighting to give a real utility feel
4-5. The powerful in-store graphics, photography and artwork 'hero-ing' the coffee bean and elements of its by-products having been beautifully selected
6. Fusion of interior and exterior

1. 店面
2、3. 裸露的砖墙和天花与水泥地面和情景照明打造了一种独特的真实感
4、5. 关于咖啡豆及其副产品的图形、照片以及艺术品
6. 室内外的融合

咖世家（Costa）连锁咖啡厅于2010年夏季在伦敦大波特兰街街角开设了第一家以都市为主题的分店。

Stiff & Trevillion事务所在2010年初参加方案竞赛，最终被选择为其设计并实施全新的理念，旨在迎合都市消费人群。咖世家一直以来备受欢迎，经过调查发现，其有机会更好的服务于都市繁华地区的消费人群，但同时面临着和其他高档餐饮品牌的竞争。

设计目标是强化品牌的特色，即"咖啡是主角"（coffee is king），并使其成为室内设计和顾客体验的亮点。原有元素如裸露的砖墙和天花与水泥地面和情景照明打造了一种独特的真实感。关于咖

啡豆及其副产品的图形、照片以及艺术品全部是由 Helen Senior 事务所精心选择、制作和设计的。

咖世家的顾客可以分为两类：一类为"速食"顾客，他们通常打包食品然后带走；另一类则为"休闲"顾客，他们在餐厅内休息放松，或者看书，或者工作，或者与朋友和同事聊天。餐厅内包括两个独立的区域，专门服务于两类顾客，营造多样性的就餐经历。

这一设计对于其连锁集团来说是非常成功的，开创了新的理念并能够广泛推广。咖世家目前在伦敦拥有 3 家分店，而在未来一年内将会开设 10 家或者更多。

Yoshinoya
吉野家

The logo of Yoshinoya of a 'bull horn was invented by Yoshinoya's founder Eikichi Matsuda and with orange as the main colour.The idea of the 'bull horn' logo derived from the initial letter of Yoshinoya's English name 'Y', while the rope surrounding the horn represents the 'Yokozuna' ranking (winner) in Japanese sumo-wrestling, also representing the 'Yokozuna' quality of the food. The surrounding rope is made up of 27 rice grains and the whole logo means Yoshinoya sells the 'best beef bowls'.

吉野家的"牛角"标志是由吉野家创办人松田荣吉设计，以"橙色"作为主色。"牛角"的标志是取自吉野家英文名称 Yoshinoya 的"Y"字而成；而围绕着"牛角"的绳状则是日本相扑中"横纲"级的代表，"横纲"亦即是相扑中的冠军，象征食品是"横纲"（即冠军）级的产品。外围那绳状由 27 粒米组成，即代表饭。因此，整个标志的意思是 Yoshinoya(吉野家)售卖"最好的牛肉饭"。

Yoshinoya Mongkok
吉野家旺角店

Completion : 2012
Location: Mongkok, Kowloon, Hong Kong, China
Designer: AS Design Service
Photographer: AS Design Service Limited
Area: 453.56m²

完成时间：2012
地点：中国 香港 九龙 旺角
设计：AS设计公司
摄影：AS设计公司
面积：453.56平方米

The biggest challenge of the project is that, having a long history, the brand's trademark and main colour is well understood by the general public, the designers tried to create a new emblem for the brand while keeping the original elements and to be innovative. This new emblem has to be simple and easy to understand, while leaving a deep impression and high malleability, varied and interesting.

The main competitors with a strong trademark in the market made essential changes to the main colours, but the client did not allow such a change, but only to add new emblem and more colours. They also considered the linkage between traditional Japanese cultures to a part of the design but not too traditional; it made the design more difficult.

Designers Four Lau and Sam Sum based on 'home' as the design concept of Yoshinoya Fast Food Restaurant, giant Japanese chopsticks, Japanese wooden plate menus and Japanese roof tiles became the feature of traditional homes and a variety of combination of diverse elements. The design of traditional Japanese chopsticks and wooden plate menus created experimental and playful design effects.

The advantages of using only 2mm thick vinyl flooring material for the curved walls and ceilings are that they can be easily bended, easily installed and low costs. For the decorations on the walls, the designers used the 'scientific Ceramic' rice bowls with specially printed traditional Japanese pattern. The colours of these bowls are durable and they are not easily broken.

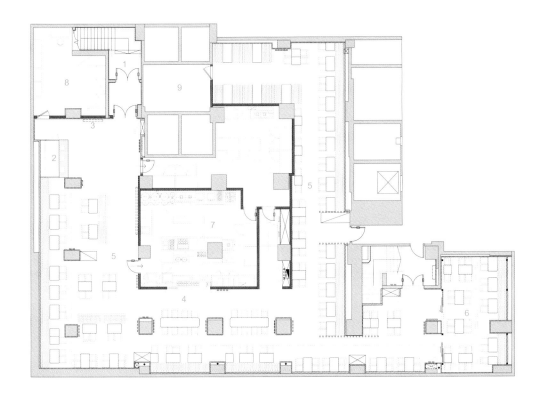

1. Entrance
2. Cashier area
3. Food menu area
4. Food counter area
5. Seating area
6. VIP seating area
7. Kitchen area
8. Crew area
9. Store room

1. 入口
2. 收银台
3. 菜单区
4. 柜台区
5. 就餐区
6. 贵宾就餐区
7. 厨房
8. 员工室
9. 存储间

Floor plan 平面图

1. Cashier with logos of restaurant
2. Rice bowls made of 'scientific ceramic' with specially printed traditional Japanese pattern
3-4. Dining area boasting homely atmosphere and showing the concept of the whole design
5. Wooden menu board highlighting Japanese style
6. The space highlighting simple yet innovative atmosphere
7. Entrance to the restaurant and words of 'chic home' on the walls connotating the style of the restaurant
8-9. The design of traditional Japanese chopsticks on the wall creating playful effects

1. 收银台凸显餐厅标识设计
2. 墙面上"饭碗"装饰图案由印着日式传统图案的瓷砖拼接而成
3、4. 就餐区内散发出家一般的氛围，展示出整个设计的主旨
5. 木质菜单板凸显日式特色
6. 空间风格简约但不失创意
7. 通往餐厅的入口，两侧墙壁上写着"现代之家"的字样，彰显出餐厅的主要风格
8、9. 墙面上巨大的日式筷子图案增添了空间趣味性

吉野家因其悠久的历史、独特的形象和典型的色彩而为大众所熟知，而这一项目的设计挑战即为：在保留原有要素的同时，打造一个全新的形象，既要简约，通俗易懂，同时又具备较强的可塑性、多样性和趣味性。

吉野家通过改变色彩而打造一个全新的品牌形象，但这一项目的客户不允许这么做。他们只许增添新的标识和色彩，并要求将传统的日式文化作为设计元素但不能过于突出传统。这无疑给设计师带来了很大的困难。

设计师以"家"为主要理念，巨大的日式筷子、日式木质菜单板、日式屋顶瓦片等传统特色被引入进来，增添空间体验感和趣味性。

弯曲的墙面和天花采用2毫米厚的乙烯材质装饰，不仅可以节约成本，更易于弯折、安装。墙面上"饭碗"装饰图案由印着日式传统图案的瓷砖拼接而成，色彩持久且不易损坏。

Yoshinoya TM
吉野家荃湾店

Location: Tsuen Wan, Kowloon, Hong Kong, China
Designer: AS Design Service Limited
Completion: 2012
Area: 337.82m²
Photographer: AS Design Service Limited

完成时间：2012
地点：中国 香港 九龙 荃湾
设计：AS设计公司
摄影：AS设计公司
面积：337.82平方米

AS Design used 'home' as the core design element to build a 'modern home' to present the brand image as young, energetic, and warm. The shape of house created a marker that makes the customers easily associated with the new image 'home'.

The cost and fast food operations mode in the design process must take into account and made the breakthrough, examples of designs that the client was satisfied are: we found that we had only change the original horizontal layout design to be vertical, the background colour with the wood grain pattern can be changed to a Japanese-style wooden menu, to create a new unique form of Japanese fast food, on the other hand they can easily be replaced at a lower production cost.

The advantages of changing all lightings in the shops (including fluorescent tubes, signs, decorative lamps, etc.) to energy saving light bulbs and LED light bulbs were long lasting and energy saving. The TV for products promotion changed to an energy-efficient LED models, it does not only reduce operational costs, but also promotes environmental protection and matches with the brand's idea of 'Quality with Conscience'.

Less is more, the simpler the more challenging, since the designers thought that a good design has to be people oriented, to gain a win-win situation between client and their customers, and to achieve both aesthetics and practicality is the most difficult, but it is also the most interesting and satisfying.

1

1. Yellow pillar structure showing the new image of Yoshionoya
2. Decorative patterns on the wall emphasising traditional Japanese style
3~5. Bright orange and calm black contrasting each other and highlighting the modern style of the space
6. Entrance to the restaurant- lighting embedded in the wall shining, leading customers in together with the logo on the ceiling

1. 黄色角状支架展现品牌新形象
2. 墙面上的装饰图案彰显传统日式风格
3~5. 艳丽的橙色和素雅的黑色形成鲜明对比，彰显空间现代感
6. 通往餐厅的入口——嵌在墙壁内的灯饰与天花上的标示共同指引着顾客进入

设计概念以"家"为中心，打造一个"现代感之家"，令企业品牌注入年轻新活力，形象更亲民。以"屋"形构思代表"家"，目的是创出一个标记令顾客容易联想到品牌新形象。

设计过程中，成本和快餐厅经营模式必须列入考虑范围之内，从而寻找突破。对于此项目，客户比较满意的几个方面如下：原有的空间格局实现了从水平方式到垂直方式的改变；背景色以及木纹图案采用日式风格木质菜单板取代；打造了一个全新而独特的日式快餐风格；利用较少的预算就可以实现空间改造。

餐厅内所有的灯饰（荧光灯管、标牌和装饰灯）采用节能灯泡和LED灯代替，满足耐用和节能的要求。用于展示餐厅产品的电视采用LED屏幕代替，不仅减少经营成本，更能推广环保理念，同时与品牌的核心"良心品质"相呼应。

在设计师看来，"少即是多"意味着越是简约，越是面临更多的挑战。他们认为，一个好的设计应该"以人为本"，实现客户和顾客的双赢。既要满足实用性，又要兼具美感是一件非常困难的事情，但同时也更充满乐趣。

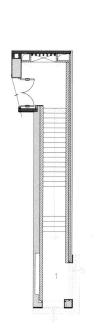
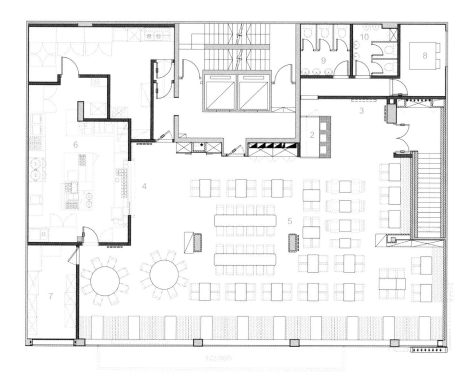

1. Entrance
2. Cashier area
3. Food menu area
4. Food counter area
5. Seating area
6. Kitchen area
7. Store room
8. AC room
9. Female toilet
10. Male toilet

1. 入口
2. 收银区
3. 菜单区
4. 柜台区
5. 就餐区
6. 厨房
7. 存储区
8. 休息区
9. 女士卫生间
10. 男士卫生间

Floor plan 平面图

Chapter 6: Technical guideline

Construction Plan Submittal Requirements

The proper layout and construction of a food facility is an important element in a successful and profitable business. It assures that you will meet all structural and operational requirements of the applicable health laws and, at the same time, meet the objective of serving safe and wholesome food to the public. The intent of this guideline is to assist you in meeting these goals.

Plans shall be easily readable, drawn to scale, (e.g. ¼" = 1') and shall include:

1. Complete floor plan with plumbing, electrical, lighting and equipment details. Demolition plans may be required for the proposed remodel of an existing food facility.

2. Complete mechanical exhaust ventilation plans including make-up air. Indicate the type of comfort cooling in the building, e.g. 'building is cooled by refrigerated air conditioning,' 'evaporative cooling,' or 'no cooling system is installed.'

3. Finish schedule for floors, walls, and ceilings that indicate the type of material, the surface finish, the colour, and the type of coved base at the floor-wall juncture.

4. A site plan including proposed waste storage receptacle location. (If applicable, specify location of common use restrooms, janitorial facilities and the On-site Management office.)

5. Equipment manufacturer's specification sheets may be required for plan review and approval.

6. Copy of the proposed menu.

7. Copy of the current Public Health Permit for the proposed remodel of an existing food facility.

施工方案要求

合理的规划与施工对于打造一个成功而盈利的食品经营机构来说至关重要。要求遵循所有有关健康法案的规定，同时确保向公众提供安全卫生的食物。准则设立的意图即为帮助经营者实现上述这些目标。

施工方案应满足易读性，并应包含以下七个方面：

1. 管道铺装、电路系统、照明设施以及设备安装细节的整体规划；拆卸计划应提供改造方案图；

2. 机械排风整体计划，包括补偿空气。其中应指明采用何种制冷方式，如空调制冷、蒸发冷却、无制冷系统等；

3. 地面、墙壁、天花的整体装饰计划，其中包括材料类型、表面抛光、色彩选择、脚线类型；

4. 废物存储空间位置确定（如果可能，指出卫生间、清洁设备及现场施工监管办公位置）；

5. 设备供应商明细单；

6. 预定菜单副本；

7. 餐厅改造公共卫生许可副本。

Construction Detail Requirements

The plans shall show and specify in the following details:

1. Full Enclosure
Each permanent food facility shall be fully enclosed in a building consisting of permanent floors, walls, and an overhead structure that meet the minimum standards as prescribed by this part. Food facilities that are not fully enclosed on all sides and that are in operation on January 1, 1985, shall not be required to meet the requirements of this section until the facility is remodelled or has a significant change in menu or its method of operation. CRFC - 114266

2. Floors
The floor surfaces in all areas (except in sales and dining areas) where food is prepared, prepackaged, or stored, employee change rooms, where any utensil is washed, where refuse or garbage is stored, where janitorial facilities are located, and in all toilet and hand washing areas, shall be smooth and of durable construction and nonabsorbent material that is easily cleanable (impervious to water, grease and acid) (e.g. quarry tile or trowelled epoxy, approved commercial grade sheet vinyl or other approved materials). Painted floor surfaces are not acceptable.

These floor surfaces shall be coved at the juncture of the floor and wall with a three-eighths inch (3/8") minimum radius coving and shall extend up the wall at least four inches (4"), except in areas where food is stored only in unopened bottles, cans, cartons, sacks, or other original shipping containers. Vinyl top set cove base is not acceptable.

Approved anti-slip floor finishes or materials are ONLY acceptable in areas where necessary for safety reasons, such as traffic areas.

Floor Drains shall be installed in floors that are water-flushed for cleaning and in areas where pressure spray methods for cleaning equipment are used. Floor surfaces in these areas shall be sloped 1:50, approximately

施工细节要求

计划应详细体现以下方面：

1. 全面围合
任何一个长期经营的食品结构都应该设立在封闭的建筑内，包括地面、墙壁以及屋顶结构。不满足上述条件的食品机构不必遵循本章提到的相关标准，另外，1985年1月1日之前成立的机构也不必遵循。但如果改变经营模式或供应的食物品种发生改变，改造时应遵循相关标准。

2. 地面
所有区域包括食品制作区、包装区、储存区、员工更衣室、餐具清洗区、垃圾处理区、清洁设备储藏区、卫生间等，其地面应满足光滑、耐用的需求，同时材料选择上符合易清洗和非吸收性要求，如大铺地砖、压光环氧树脂等。一定要避免喷漆地面。

上述这些地面材质在脚线处要形成最小半径为3/8英寸的铺设范围，并延伸到墙面4英寸。储存灌装食品的区域除外，脚线处不能采用塑料材质。

防滑地面装饰或材质仅在满足安全需求的区域内使用。

地面排水管道应安装在需冲水清洗的地面处和可使用压力喷雾清洗设备的区域内。这些区域的地面应形成1:50的坡度，大约以每平方英尺1/4英寸的标准向排水管道倾斜。如果地面没有坡度，那么排水管道四周两英尺之内应增添一定坡度。

Chapter 6: Technical guideline

one-quarter inch (¼") per foot toward the floor drains. When floor drains are added to an existing facility where the floor surface is not sloped, a two (2) foot surrounding depression/slope to the floor drain may be required.

3. Wall and Ceilings
The walls and ceilings of all rooms shall be of a durable, smooth, non-absorbent, easily cleanable surface except in the following areas: (a) bar areas in which alcoholic beverages are sold or served directly to the consumers, except wall areas adjacent to bar sinks and areas where food is prepared; (b) areas where food is stored only in unopened bottles, cans, cartons, sacks, or other original shipping containers; (c) dining and sales areas; (d) offices; (e) restrooms used exclusively by the consumers, except that the walls and ceilings shall be of a nonabsorbent and washable surface. Acceptable materials are gloss or semi-gloss enamel paint, epoxy paint, FRP (Fiberglass Reinforced Panel), stainless steel, ceramic tile or other approved materials and finishes. Acoustical ceiling tile may be used if it is installed not less than six feet above the floor. Exposed brick, concrete block, rough concrete, rough plaster or textured gypsum boards are not acceptable. A sample may be required for review. Walls and ceilings of food preparation areas, restrooms, janitorial areas, utensil washing areas, and interior surfaces of walk-in refrigeration units are recommended to be light coloured. Light colour shall mean having a light reflectance value of 70 percent or greater.

Conduits of all types shall be installed within walls as practicable. When otherwise installed, they shall be mounted or enclosed so as to facilitate cleaning (e.g., at least ½ inch from the wall and six (6) inches above the floor surface).

4. Lighting
In every room and area in which food is prepared, processed or packaged, or in which utensils are cleaned, sufficient natural or artificial lighting shall be provided to produce the following light intensity while the area is in use:

3. 墙面和天花
墙面和天花表面应满足耐用、光滑、非吸收及易清洗要求。但以下区域除外：（a）酒类饮品售卖区；(b)灌装食品储存区；(c)就餐区和销售区；（d）办公区；（e）顾客专用卫生间。可选择的材料包括上釉油漆、压光油漆、纤维玻璃板、不锈钢板、瓷砖或其他认证可使用的材料。天花在地面英寸之上可使用吸音瓷砖。裸露砖材、混凝土砌块、未经修饰的混凝土、石膏等不可使用。

食品制作区、卫生间、设备清洁区以及步入式冷藏区内部的墙面及天花推荐使用浅色系装饰。其中，浅色系是指光反射值应达到70%或以上。
管线应安装在墙壁内，并使用特殊材质包裹起来，以便易于清洗。（与墙面的距离至少保持在1/2英寸，与地面的距离至少为6英寸）

4. 照明
在每个房间和区域如食物制作、加工、包装区及设备清洗区，应保证足够的自然或人造光线的供应，不同区域的光线强度要求如下：

a) At least 10-foot candles at a distance of 30 inches above the floor in walk-in refrigeration units and dry food storage areas.
b) At least 20-foot candles where food is provided for consumer self-service; where fresh produce or prepackaged foods are sold or offered for consumption; inside equipment such as reach-in and under-counter refrigerators; handwashing areas; warewashing areas; equipment and utensil storage areas; and in toilet rooms.
c) At least 50-foot candles at surfaces where a food employee is working with food or with utensils, equipment such as knives, slicers, grinders, or saws where employee safety is a factor; and in other rooms during periods of cleaning.

Light bulbs shall be shielded, coated, or otherwise shatter-resistant in areas where there are non-prepackaged ready-to-eat foods, clean equipment, utensils, and linens, or unwrapped single-use articles.

5. Ventilation
Ventilation shall be provided to remove toxic gases, odour, steam, heat, grease, vapours, or smoke from the food facility including: food preparation, scullery, toilet, janitorial, garbage and change rooms. All areas of a food facility shall have sufficient ventilation to facilitate proper food storage and to provide a reasonable condition of comfort for each employee, consistent with the job performed by the employee.

Mechanical Exhaust Ventilation shall be provided over all cooking equipment such as ranges, multiple table top cooking units, broilers, fry grills, steam jacketed kettles, griddles, ovens, deep fat fryers, barbecues, rotisseries, high temperature warewash machines (160°F rinse water), and similar equipment to effectively remove steam, heat, grease, vapours, cooking odours and smoke from the food facility. Generally, chemical sanitizing or under-counter warewash machines do not require exhaust hoods. Ventilation plans for each system shall include front and side elevation of the exhaust hood and duct details to the roof fans (both exhaust and make-up air). Provide manufacturer specification sheets for exhaust fan, make-up air fan and hood filters along with the

a) 步入式冷藏区及干燥食品存储区内，至少每隔30英寸应安装规格为10英尺的照明灯；
b) 自助餐区、新鲜食物或包装食物售卖区、步入式冷藏区、洗手区、设备清洗区、设备储藏区及卫生间内至少安装规格为20英尺的照明灯；
c) 工作人员利用刀、叉等工具制作食物的区域应安装规格为50英尺的照明灯。

未经包装的食品、清洁设备、织品工具或没有包装的一次性用品区内，使用灯泡照明应将其遮盖或包裹起来。

5. 通风
食品制作区、后厨房、卫生间、清洁室、垃圾处理区及更衣室内应确安装通风设备，以便于除去有毒气体、气味、蒸汽、热气、油脂、油烟等。所有区域内都应保证通风良好，便于保存食物并为员工提供舒适的工作环境。

机械排风应满足相关标准。所有烹饪设备包括炉灶、桌面烹饪设备、烤架、蒸汽锅、圆烤盘、烤箱、深油炸锅、电转烤肉架、高压清洗机以及其他一些除油烟、蒸汽等的设备上都应安装排风设备。一般情况下，化学清洗设备等上方不需排风。每个系统的排风计划都应包含排风设备的前后立面以及连接屋顶风扇的管道细节处理。提供供应商明细单，包括排风扇、空气补给扇、排烟过滤器以及静压测算值。可供参考的标准有商业厨房排废系统统一机械规范、商业食品加工设备机械排风及油烟系统准则等。同时确定空气补给分散器的数量和位置，在最终确定之前应提交第三方出具的空气平衡报告。

Chapter 6: Technical guideline

static pressure calculations. Refer to the Uniform Mechanical Code Chapter on Commercial Kitchen Ventilation Systems. Refer to CCDEH Recommendations for Mechanical Exhaust Ventilation and Hood Systems for Commercial Food and Utensil Heat Processing Equipment Guidelines. Specify the number and location(s) of make-up air diffusers. An air balance report completed by a third party may be required prior to final clearance.

Toilet rooms shall be vented to the outside air by means of an openable, screened window, an air shaft, or a light-switch activated exhaust fan consistent with the requirements of local building codes.

6. Flies, Rodent and Vermin Exclusion
A food facility shall at all times be constructed, equipped, maintained, and operated to prevent the entrance and harbourage of animals, birds and vermin, including, but not limited to, rodents and insects.

Windows: All openable windows shall be provided with approved screening not less than 16 mesh per square inch set in tight fitting frames. Pass-through window service openings shall be limited to 216 square inches each. Each opening shall be provided with a solid or screened window, equipped with a self-closing device. Screening shall be at least 16 mesh per square inch. Pass-through windows up to 432 square inches are approved if equipped with an air curtain device. Window openings must be closed when not in use. The minimum distance between the openings shall not be less than eighteen inches (18").

Delivery Doors: All delivery doors leading to the outside shall be self-closing. Overhead, automatic switch-activated air curtains must be provided at delivery doors. The air curtain will produce an air flow eight (8) inches thick at the discharge opening and with an air velocity of not less than 1600 FPM (feet per minute) across the entire opening measured at a point three (3) feet below the air curtain. Food facilities that sell only pre-packaged food are the only food facilities may be exempted from this requirement. Large cargo-type doors shall not open directly into a food preparation or utensil washing area.

卫生间排风应通向室外，借助可开闭的窗户、通风井及可控排风扇等设备，同时遵循当地建筑法则规定。

6. 苍蝇、啮齿动物及害虫的防范
食品经营结构在任何情况下都应该保证阻止各种动物包括鸟类和害虫等进入，更应确保不能成为它们的避难所。

窗户：所有可开闭的窗户遮阳屏应满足每平方英寸不超过 16 网眼，并确保镶嵌在结实的框架内。传递窗开口大小应不超过 216 平方米英寸，并采用屏幕遮挡，其大小如达 432 平方米英寸则应采用空气幕遮挡。另外，开口之前的距离不超过 18 英寸，而且在不使用时应保持关闭。

门：所有通向室外的门须可自动关闭，而且配备自动空气帘。空气帘在开启时可产生 8 英寸厚的空气流，空气流速不小于 1600 英尺/每分钟（这一数据是在空气帘下 3 英尺的距离内测算出来的）。仅出售包装食物的经营机构可以不必遵循上述准则。另外，大型运货门不应直接朝向食物制作区或餐具清洗区开闭。

Exterior Doors: All exterior doors leading to the outside shall be self-closing, tight-fitting and vermin proof. Air curtains may be used as auxiliary fly control but are not adequate substitute devices to permit a door to remain open.

Vector Control: Openings at the base and side of exterior doors shall not exceed one-fourth inch (¼"). All exterior wall pipes or other openings shall be tightly sealed. All exterior wall vents shall be properly screened with one-fourth inch (¼") hardware cloth screening.

Garbage and Trash Area: Each food facility shall be provided with facilities and equipment necessary to store or dispose of all waste material. An area designated for refuse, recyclables, returnables, and a redeeming machine for recyclables or returnables shall be located so that it is separate from food, equipment, utensils, linens, and single-service and single-use articles and a public health hazard or nuisance is not created. If located within the food facility, a storage area for refuse, recyclables, and returnables shall meet the requirements for floors, walls, ceilings, and vermin exclusion. If provided, an outdoor storage area or enclosure used for refuse, recyclables, and returnables shall be constructed of nonabsorbent material such as concrete or asphalt and shall be easily cleanable, durable, and sloped to drain.

7. Toilet Facilities
 Each permanent food facility shall be provided with clean toilet rooms, in good repair for the use by employees. The number of toilet and handicapped facilities required shall be in accordance with local building and plumbing ordinances. Toilet rooms shall be separated from other portions of the food facility by well-fitted, self-closing doors. Toilet tissue shall be provided in a permanently installed dispenser at each toilet. CRFC – 114250, 114276

Toilet facilities (at least one separate toilet facility for men and one separate toilet facility for women) in good repair shall be provided for consumers, guests, or invitees, when the food facility was constructed

室外门：所有通向室外的门须可自动关闭、封闭性强，避免动物进入。空气帘可用作阻挡苍蝇等物体进入，但不是必须元素。

开口控制：室外门下部和侧面的开口不能超过 1/4 英寸；室外墙壁管道和开口应严格密封；室外墙壁通风口采用 1/4 英寸金属布遮挡。

垃圾废料处理区：所有食物经营机构都应该配备用于存储和处理垃圾的空间和设备。用于存储和处理废料的空间应远离食物、厨房设备、餐具以及一次性使用的物件，以确保卫生安全。入口存储区设于食物经营机构内，其设计应满足上面曾提到的相关准则。如设于室外，应采用非吸收材料如混凝土或沥青建造，确保易于清洗、持久耐用并满足排水坡度要求。

7. 卫生间设备
每一家长期的食物经营机构都应该提供整洁的卫生间。其中，卫生间的数量及无障碍设备应尊循当地建筑法规。卫生间与其他空间隔开，卫浴设备附近提供卫生纸巾。

机构建于 1984 年 7 月 1 日之后，建筑面积超过 20000 平方英尺，应为顾客等提供单独的男士和女士卫生间。

Chapter 6: Technical guideline

after July 1, 1984, and has more than 20,000 square feet of floor space. A any building that is constructed after January 1, 2004, that provides space for the consumption of food on the premises shall provide clean toilet facilities in good repair for consumers, guests or invitees. These facilities shall be located where consumers, guests, and invitees do not pass through food preparation, food storage, or utensil washing areas to reach the toilet facilities.

Food facilities located within amusement parks, stadiums, arenas, food courts, fairgrounds, and similar premises shall not be required to provide toilet facilities for employee use within each food facility if the following conditions are met:

a) Approved common use toilet facilities are located within 200 feet in travel distance of each food facility.
b) Approved common use toilet facilities are readily available for use by employees.
c) The On-site Management office shall be staffed with personnel responsible for the maintenance of the designated common toilet facilities.
d) Prior approval must be obtained from the Health Department Plan Check.
e) Approved common use toilet facilities shall be located in a common area that will not be accessible through another business.

Handwashing facilities shall be provided within or adjacent to toilet rooms and shall be equipped with an adequate supply of hot and cold running water under pressure. Handwashing sinks shall have water provided from a combination or premixing faucet which supplies warm water (at least 100°Farenheit) for a minimum of fifteen (15) seconds while both hands are free for washing. The number of handwashing facilities required shall be in accordance with local building and plumbing ordinances. Handwashing cleanser and single-use sanitary towels or hot-air blowers shall be provided in dispensers at all handwashing facilities.

凡是建于 2004 年 7 月 1 日之后的食品经营机构都应该配备整洁的卫生间，供顾客等使用。卫生间不应设置在食品制作区、存储区及餐具清洗区内。

食品经营结构如设立在公园、体育场、竞技场、美食广场、市场等其他场所内。在满足如下条件的前提下，不要求每家机构都为员工提供卫生间：

a) 周围 200 英尺内设有公共卫生间；
b) 公共卫生间可供员工使用；
c) 管理办公室应配备专门人员负责卫生间的维护与修理；
d) 经健康部门检查合格；
e) 公共卫生间应设置在公共区域内，不为场所之外的其他机构服务。

卫生间内或附近区域应配备洗手设备，并确保足够的冷热水供应。洗手盆应安装混合水龙头，并确保在 15 秒钟之内供应最少 100 华氏温度的热水。洗手设备的数量应满足当地建筑准则要求，设备附近应供应洗手清洁剂及一次性使用的卫生毛巾或热气吹风机。

Where alcoholic beverages are sold or given away for consumption on the premises there shall be provided for use by the public, separate toilet rooms for each gender, with the men's toilet room having at least one urinal. At least one lavatory shall be provided in conjunction with and convenient to each toilet room. The toilet rooms must be located within the food facility and where consumers, guests, and invitees do not pass through food preparation, storage, or utensil washing areas to reach the toilet facilities.

8. Employee Changing Room

Lockers or other suitable facilities shall be located in a designated room or area where contamination of food, equipment, utensils, linens, and single-use articles cannot occur. Dressing rooms or dressing areas shall be provided and used by employees if the employees regularly change their clothes in the facility.

9. Equipment Standard

All equipment shall be designed and constructed to be durable and to retain their characteristic qualities under normal use conditions. All new and replacement food-related and utensil-related equipment shall be certified or classified for sanitation by an American National Standards Institute (ANSI) accredited certification programme. In the absence of an applicable sanitation certification, unique or special equipment may be evaluated for approval by the local enforcement agency. All materials that are used in the construction of utensils and food contact surfaces of equipment shall not allow the migration of deleterious substances or impart colours, odours, or tastes to food; and under normal use conditions shall be safe, durable, corrosion-resistant, nonabsorbent, sufficient in weight and thickness to withstand repeated washing, finished to have a smooth, easily cleanable surface, and resistant to pitting, chipping, crazing, scratching, scoring, distortion and decomposition.

Handwashing Sink: Food facilities constructed or extensively remodeled after January 1, 1996, that handle non-prepackaged food, shall provide facilities exclusively for handwashing in the food preparation areas and

酒类饮品售卖区内应配备独立的男士及女士卫生间，其中男士卫生间内必须至少设置一个小便池。卫生间不应设置在食品制作区、存储区及餐具清洗区内。

8. 员工更衣室

指定的区域内应配备储物柜或其他合适的设备，但要远离食物、餐具或一次性使用的物品。如果员工需要经常更换衣服，则应设置专门的更衣室或更衣区。

9. 设备标准

所有的设备都应满足耐用性，并确保在正常使用的条件下不会受损。所有新添和替换的同食物或餐具相关的设备都应经美国标准协会认证。如缺乏卫生认证，一些特殊的设备引进应经当地执行机构批准。用于制作餐具的材料或与食物直接接触的材料必须禁止有害物质、颜色或气味进入到食物之内。在正常使用情况下，应满足安全、耐用、防腐蚀、非吸收、耐磨、易清洗等特征。

洗手盆：1996年1月1日之后成立的或改建的从事非包装类食品售卖的食品经营机构都应该在食品制作区及其之外的洗浴区配备专门洗手盆。洗手设备在数量和方便性上应满足一定的要求，并保持整洁，供店内员工随时使用。手盆应安装混合水龙头，并确保在15秒钟之内供应最少100华氏温度的热水。设备附近应供应洗手清洁剂及一次性使用的卫生毛巾或热气吹风机。

Chapter 6: Technical guideline

in warewashing areas that are not located within or immediately adjacent to food preparation areas. Handwashing facilities shall be sufficient in number and conveniently located, maintained clean, unobstructed and accessible at all times for use by food employees. Handwashing sinks shall have water provided from a combination or premixing faucet which supplies warm water (at least 100°Farenheit) for a minimum of fifteen (15) seconds while both hands are free for washing. Handwashing facilities shall be provided with handwashing cleanser and sanitary single-use towels.

Food Preparation Sink: Food facilities are required to have a separate sink for when they are engaged in activities such as washing, rinsing, soaking, thawing, or similar preparation of foods, and shall be located within the food preparation area. The sink shall have an integral drainboard, and have minimum tub dimensions of 18" x 18" by 12" deep. An adjacent work table of similar dimensions may be substituted for the drainboard. A food preparation sink must drain indirectly through an air gap into a floor sink and must be free standing (not installed in cabinets).

Manual Warewashing Sink: All food facilities, except those that have only prepackaged items in their original unopened sealed containers, shall provide a three-compartment warewashing sink with two integral drainboards. The tub compartments shall be large enough to accommodate immersion of the largest piece of equipment and utensils; i.e., minimum 16" x 20" or 18" x 18" by 12" deep. The sink must be free standing (not installed in cabinets). Hot and cold running water under pressure shall be provided to each compartment. When the three-compartment sink is installed next to a wall, a metal 'back splash' shall extend up the wall a minimum of eight (8) inches, and shall be formed as an integral part of the unit and sealed to the wall. The manual warewashing sink shall be easily accessible and conveniently located to the food preparation area.

Bar Sink (Warewashing): A three (3) compartment bar sink (minimum 10" X 14" X 12") with two integral metal drainboards (minimum 18" x 14") shall be provided in bars.

食品准备清洗池：食品经营机构在食品制作区应配备独立的清洗池用于清洗、冲洗、浸泡或解冻原料。清洗池内应设有整体滴净板，最小规格18x18x12（英寸）。同样规格的操作台可代替滴净板，清洗池排水应经由空气隙流向地面排水管，而且必须独立存在。

人工器皿洗涤槽：所有的食品经营机构（除只经营包装食品的机构）应该配备带有两个整体滴净板的三格洗涤槽，其必须是独立的，规格应满足能容纳最大的餐具和设备，例如，最小规格应满足16x20x16（英寸）或18x18x12(英寸)。确保每格内都有冷热水供应。如果洗涤槽安装在墙面上，金属连壁应向墙面延伸至少8英寸，并与墙面牢固地接合在一起。洗涤槽应确保方便使用。

吧台水槽：酒吧区内应配备带有两个整体金属滴净板的三格水槽（最小规格为10x14x12英寸），其中滴净板最小规格应为18x14英寸。

Warewashing Machines: Mechanical warewashing shall be accomplished by using an approved machine installed and operated in accordance with the manufacturer's specifications. Warewashing machines shall have two integral drainboards that are of adequate size and construction to accommodate the warewashing trays (i.e., 24" x 24").

The drain-boards shall be sloped and drained to an approved waste receptor. Where an under-counter warewashing machine is used, there shall be two metal drainboards located adjacent to the machine. This requirement may be satisfied by using the drainboards that are part of the manual warewashing sinks if the facilities are located adjacent to the machine.

Pot and pan washers shall be equipped with drainboards or shall be equipped with approved alternative equipment that provides adequate and suitable space for soiled and clean equipment and utensils.

NOTE: Installation of a mechanical warewashing machine does not eliminate the requirement for a 3-compartment sink.

Rinse/work sink: A single compartment rinse/work sink may be provided in service areas where blenders or similar equipment are rinsed and the three-compartment sink is not located within the area. NOTE: Rinse/work sink will not be a substitute for the requirement of a 3-compartment sink.
Dump sink: A single compartment dump sink is generally installed in alcoholic beverage bars for the emptying of the contents of used pitchers and glasses.

Drainage connection for warewashing, rinse/work sink, dump sink and similar type equipment: Verify with the local building department for the proper connection of the drainage system to sewer. A direct connection will result in the requirement of an adjacent (within five feet) floor drain upstream of the warewashing equipment, as per the Uniform Plumbing Code Section 704.3.

洗碗机：机械清洗设备应经认证，并按照生产商要求使用。洗碗机应配备两个规格足够大的整体滴净板，其中滴净板应倾斜放置，使水流到固定容器内。放在厨房台面下的洗碗机，应在附近配备两个金属滴净板。如果洗碗机放置在器皿洗涤槽附近，则可与其共用滴净板。

锅壶清洗机应配备滴净板或者其他可放置干净设备和器皿的空间。

注意：安装洗碗机并不意味着可以取代三格洗涤槽。

漂清池：应为沙冰机或其他需冲洗设备配备单格漂清池，三格洗涤槽不应放在相同区域内。

注意：漂清池不能代替三格洗涤槽。

废物清洗池：酒吧或饮料吧应配备单格废物清洗池，用于清空用过的杯子或瓶子等容器内的废物。

器皿洗涤槽、漂清池、废物清洗池等的排水连接：根据当地建筑部门对于排水系统与下水道连通的要求确定，直接连接将会导致地面排水附近的设备位于清洗设备上游。

Chapter 6: Technical guideline

Janitorial Sink: At least one (1) of the following is to be used for general cleaning purposes and for the disposal of mop bucket wastes and other liquid wastes:

a) A one-compartment, non-porous janitorial sink.
b) A slab, basin, or curbed cleaning facility constructed of concrete or equivalent material, and sloped to a drain.

Such facilities shall be connected to an approved sewerage, and provided with hot and cold running water through a mixing valve and protected with a backflow protection device. Janitorial sinks and basins shall be separated from other equipment by at least 30 inches; or a solid partition that extends a minimum of 18 inches above the rim or top of the janitorial sink or basin.

A room, area, or cabinet separated from any food preparation, food storage area or warewashing area shall be provided for the storage of cleaning equipment and supplies, such as mops, buckets, brooms, cleansers and waxes. CRFC – 114279, 114281

Janitorial facilities shall not be required within each food facility if the following conditions are met:

a) Approved common use janitorial facilities are located within 100 feet in travel distance of each food facility and shall be on same level as the food facility.
b) Approved common use janitorial facilities are readily available for use by employees.
c) The On-site Management office shall be staffed with personnel responsible for the maintenance of the designated common use janitorial facilities.
d) Prior approval must be obtained from the Health Department Plan Check.
e) Approved common use janitorial facilities shall be located in a common area that will not be accessible through another business.

清洁池：以下提到的必须满足一项，用于通用清洗或冲洗拖把或其他液体废物。

a) 单格、非多孔构造的清洁池；
b) 由水泥等材质打造的水池，并朝向排水管倾斜；此类清洁设备应与下水道连通，通过混合阀门同时提供冷、热水并保护防止逆流。清洁池与水盆必须间隔开，最小间距为30英寸或者采用固体隔开，超过水池或水盆上边缘至少18英寸。

应配备单独的房间、区域或橱柜用于存放清洁设备，如拖布、桶、扫把、清洁剂等。但确定这些区域应与食物制作区、存储区和设备清洗区分隔开来。

如果满足以下条件，则不必配备清洁设备：

a) 100英尺之内设有公共清洁设备；
b) 公共清洁设备方便员工使用；
c) 管理办公室应配备专门人员负责清洁设备的维护与修理；
d) 经健康部门检查合格；
e) 清洁设备应设置在公共区域内，不为场所之外的其他机构服务。

Food and Equipment Protection: Non-prepackaged food on display and food contact surfaces shall be protected from contamination by the use of packaging, counter, service line, or sneeze guards that intercept a direct line between the consumer's mouth and the food being displayed, containers with display cases, mechanical dispensers, or other effective means. Cafeteria, buffet and salad bar self-service, food preparation equipment and food preparation areas shall be protected by approved sneeze guards. Non-prepackaged food may be displayed and sold in bulk on other than self-service containers if the food is served by a food employee directly to a consumer.

Equipment for cooling and heating food and for holding cold and hot food shall be sufficient in number and capacity to ensure proper food temperature control during transportation and operation.

Garbage Disposals: Garbage disposals may be installed in drainboards if the drainboard is lengthened to accommodate the disposal cone in addition to the minimum required drainboard size. Garbage disposals may not be installed under a sink compartment, unless an additional compartment is provided for the disposal. Verify with local building department requirements prior to installation of a garbage disposal.

Reach-In Refrigeration: Reach in refrigerators shall either be self-contained or drained indirectly to a floor sink. Refrigeration units shall be provided with an accurate, readily visible thermometer, have shelving that is nonabsorbent, non-corrodible, easily cleanable, and shall meet all applicable NSF/ANSI standards. Wood is not an acceptable structural material finish of the refrigeration unit.

食物和设备防护：未包装食物和食物接触表面应采用包装、柜台、喷嚏挡或其他有效方式间隔，避免顾客与食物直接接触。自助餐厅内，食物制作设备和制作区应采用喷嚏挡隔开。如果食物制作之后直接卖给顾客，未包装食物可以成批量展示或售卖。

确保配备一定数量和容积的、用于制冷、加热或保温的设备，确保食物运输和经营过程中能够保持适当的温度。

垃圾处理：如滴净板在满足基本规格要求并可延长，垃圾处理装置应安装在滴净板处。除非具备额外的用于垃圾处理的水槽，否则，垃圾处理装置不可安装在水槽格下方。垃圾处理装置安装之前，应满足当地建筑相关标准。

普通冷柜（卧式或立式）：普通冷柜或是自带排水系统或是连接到地面排水管道。冷柜应配备精确且可见的温度表以及非吸收、易清洗、不易腐蚀的隔断空间，并满足美国国家标准协会认证。木材不可用于冷藏柜的构造及装饰材料。

Index 索引

Ab Rogers Design
Address: 22 Parkside, London, SW19 5NA, UK
Tel: 020 8944 7088
Fax: 020 8944 1754

AS Design Service Limited
Address: Rm B1, 7/F Yeung Yiu Chung(6) Ind's Bldg,
19 Cheung Shun St, Lai Chi kok, HK
Tel: 852 2191 6433
Fax: 852 2191 6422

Baynes & Co
Address: 17 church street, Long Buckby, Tajikistan
Tel: 013 2784 4927

Blacksheep
Address: 13-19 Vine Hill, London, UK, EC1R 5DW
Tel: 44 (0) 20 7713 7413
Buckley Grey Yeoman
Address: Studio 4.04, The Tea Building, 56 Shoreditch High Street,
London E1 6JJ
Tel: 020 7033 9913
Fax: 020 7033 9914

Checkland Kindleysides
Address: 302 Clerkenwell Workshops, 27/31 Clerkenwell Close
London EC1R 0AT
Tel: 44 (0)116 2644 700

Design Clarity
Address: Studio 204 61 Marlborough St Surry Hills NSW,
2010 Australia
Tel: 61 (0)2 9319 0933

Heyroth & Kürbitz freie Architekten BDA
Address: Shanghaiallee 10, 20457 Hamburg, Germany
Tel: 040 3750 3683
Fax: 040 3750 3684

Jonathan Clark Architects
Address: 3rd Floor 34-35 Great Sutton Street
London EC1V 0DX
Tel: 020 7608 1111

José Orrego
Address: Nueva Dirección Calle Boulevard 162 Oficina 501
(espalda edificio CRONOS) Santiago de Surco, Lima - Perú
Tel: 511 4375635

Luca Bertacchi & Sara Bergami
Address: via Santo Stefano – 168,
40125 – Bologna – Italy
Tel: 39 0515873070
Fax: 39 3384743658

Outofstock Design
Address: Singapore 23 Hythe Road
Tel: 0065 94233149

Stiff + Trevillion
Address: 16 Woodfield Road London, W9 2BE
Tel: 020 8960 5550
Fax: 020 8969 8668

Studio Patrick Norguet
Address: 38 Rue de Malte 75011 Paris, France
Tel: 33 (0)1 4807 2995

Viereck Architects Ltd.
Address: Grazerstrasse 24 8650 Kindberg
Tel: 43 0 3865 3361 10

图书在版编目（CIP）数据

连锁餐厅 /（美）克利弗编；鄢格译. -- 沈阳：辽宁科学技术出版社，2013.7
ISBN 978-7-5381-8045-9

Ⅰ．①连… Ⅱ．①克… ②鄢… Ⅲ．①餐厅－室内装饰设计－图集 Ⅳ．①TU241-64

中国版本图书馆CIP数据核字(2013)第095857号

出版发行：辽宁科学技术出版社
（地址：沈阳市和平区十一纬路29号　邮编：110003）
印　刷　者：利丰雅高印刷（深圳）有限公司
经　销　者：各地新华书店
幅面尺寸：215mm×285mm
印　　张：14
插　　页：4
字　　数：50千字
印　　数：1～1800
出版时间：2013年 7 月第 1 版
印刷时间：2013年 7 月第 1 次印刷
责任编辑：陈慈良　鄢　格
封面设计：何　萍
版式设计：何　萍
责任校对：周　文
书　　号：ISBN 978-7-5381-8045-9
定　　价：228.00元

联系电话：024-23284360
邮购热线：024-23284502
E-mail: lnkjc@126.com
http://www.lnkj.com.cn
本书网址：www.lnkj.cn/uri.sh/8045